# Inorganic Chemical Nomenclature

# Inorganic Chemical Nomenclature

## Principles and Practice

B. Peter Block
Warren H. Powell
W. Conard Fernelius

American Chemical Society, Washington, DC 1990

Library of Congress Cataloging-in-Publication Data

Block, B. Peter, 1923–

Inorganic chemical nomenclature : principles and practice
B. Peter Block, Warren H. Powell, W. Conard Fernelius

p. cm.

Includes bibliographical references and index.

ISBN 0–8412–1697–5 (cloth). ISBN 0–8412–1698–3 (paper).

1. Chemistry, Inorganic—Nomenclature.

I. Powell, Warren H., 1934– . II. Fernelius, W. Conard III. Title.

QD149.B59 1990
549'.014—dc20 90–760
CIP

The paper used in this publication meets the minimum requirements of American National Standard for Information Sciences—Permanence of Paper for Printed Library Materials, ANSI Z39.48–1984. ∞



PRINTED IN THE UNITED STATES OF AMERICA

# About the Authors

B. PETER BLOCK received his B.S. degree in chemistry from Harvard, his M.S degree in organic chemistry from the University of Tennessee, and his Ph.D. in inorganic chemistry from the University of Illinois. He also did graduate work at Purdue University. After serving on the faculties of the University of Chicago and the Pennsylvania State University for 8 years, he joined Pennwalt Corporation's R&D staff. He retired from Pennwalt as a senior research scientist after over 28 years with the company and is currently an adjunct professor in the chemistry department at East Tennessee State University.

He has authored or coauthored more than 60 papers, written review articles or chapters for several books and encyclopedias, and given a number of invited lectures on his work. His research interests have centered on the synthesis and characterization of coordination compounds, with emphasis on polymeric metal phosphinates. In 1968 he was given the ACS Philadelphia Section Award for his pioneering discoveries in the field of inorganic chemistry, with particular emphasis on the preparation, stability, and structure of coordination compounds and double-bridged linear inorganic polymers. He and his co-workers hold more than 20 patents on inorganic polymers. For 12 years he was also deeply involved in the chlorofluorocarbon–ozone issue, much of that time as the Pennwalt member of the Chemical Manufacturers Association Fluorocarbon Program Panel.

Block has served on the nomenclature committees of the American Chemical Society, the ACS Division of Inorganic Chemistry, the ACS Division of Polymer Chemistry, Inc., and the National Research Council. He has contributed to the development of various nomenclature procedures and played a major role in devising the current methods for naming inorganic polymers.

WARREN H. POWELL received a B.S. degree in chemistry from Antioch College and a Ph.D. degree in organic chemistry from The Ohio State University, where he served as an assistant instructor and instructor in the department of chemistry. After 5 years as a research chemist in the organic chemicals department of E.I. du Pont de Nemours & Co. in Wilmington, Delaware, he joined Chemical Abstracts Service in 1964 and has been a senior editor since 1973.

He has participated in the activities of the nomenclature committees of the American Chemical Society and its Division of Inorganic Chemistry. He has served as associate and titular member of the IUPAC Commission on Nomenclature of Inorganic Chemistry, as associate and titular member and secretary of the IUPAC Commission on Nomenclature of Organic Chemistry, and as associate member of the IUPAC Interdivisional Committee on Nomenclature and Symbols, representing the Organic Chemistry Division.

Powell has given some 20 presentations and publications, almost all of which have been on various aspects of nomenclature. He has been directly involved in the development of a wide range of IUPAC and ACS nomenclature-related projects, including the subgroups of the elements in the periodic table, names for the super-heavy elements, nitrogen hydrides, isotopically modified inorganic and organic compounds, linear inorganic and organic single-strand polymers, polyboron hydrides, nomenclature for rings and ring systems, particularly the revision of the Hantzsch–Widman nomenclature system for heteromonocycles, and the development of the lambda and delta conventions for designating nonstandard valences and cumulative double bonds in ring names.

W. CONARD FERNELIUS received A.B., A.M., and Ph.D. degrees in inorganic chemistry from Stanford University and honorary Sc.D. degrees from Franklin and Marshall College and Kent State University. He served on the faculties of the Ohio State University, Purdue University, Syracuse University, Pennsylvania State University, University of South Florida, and Kent State University. In addition to his academic work, he was a director on the Manhattan Project and associate director of research for the Koppers Company. He received a Guggenheim fellowship at Oxford University and a Fulbright fellowship at the University of Cairo, the Pittsburgh Award in Chemistry from the Pittsburgh Section, and the Patterson–Crane Award from the Dayton and Columbus Sections of the American Chemical Society.

His research activities included studies on nitrogen systems of compounds, liquid ammonia solutions, coordination compounds, radioactivity, less-known elements, and electron spectroscopy.

Fernelius participated in the founding of *Inorganic Syntheses* and was a member of the editorial boards of several publications. He wrote five books, published over 200 articles, and edited "Real World of Industrial Chemistry" and coauthored "Notes on Nomenclature" for the *Journal of Chemical Education*. As a member of the IUPAC Commission on Nomenclature of Inorganic Chemistry for over 20 years, he was a major contributor to the 1970 edition of the IUPAC *Nomenclature of Inorganic Chemistry*, and, while chairman (1971–1975), almost single-handedly produced *How to Name an Inorganic Substance: A Guide to the Use of Nomenclature of Inorganic Chemistry, Definitive Rules, 1970* (Pergamon Press, 1977). He also served on committees on documentation and nomenclature of the National Research Council and the American Chemical Society. At the time of his death he was still actively involved in the ACS–IUPAC attempts to resolve the confusion in naming the subgroups of elements in the periodic table.

# Contents

# Preface

This book is concerned with the nomenclature of all substances except those that have carbon as a central atom and contain carbon-to-carbon bonds. Its purpose is to provide the reader with a basic understanding of the principles currently in use for naming such substances and the background necessary for searching much of the older literature. Although the emphasis is on names for individual substances, names of other sorts are also important in inorganic chemistry. They include names for groups of elements, for types of reactions, for characteristic properties, etc. Some such names are discussed; however, the systematization of such names has not progressed nearly as far as that for the nomenclature of individual chemical species.

Chemical nomenclature is used for many purposes by many people, not all of whom are chemists. Therefore, it must be capable of being understood by a great variety of people. In some situations, the only requirement of a name is that it identify a substance. When all participants are familiar with local terminology, the use of short, nonsystematic names or code designations is acceptable. It is preferable in a general audience, however, to use names that describe composition, structure, or both, so that the subject matter can be better understood.

Accurate recording of information requires that the names used in primary publications leave no question whatsoever as to the identity of the substances involved. Systematic nomenclature methods make this task much easier for both author and reader. Economy and convenience in the storage and retrieval of information dictate that the nomenclature used in indexes and computer data bases should be so restrictive that there is one and only one name for each and every substance. Rules for such a purpose must be much more extensive and sharply defined than those for general nomenclature. For the latter it is necessary only that a name for a substance be unambiguous.

The primary objective of inorganic nomenclature, like that of organic nomenclature, is to provide a 1:1 correspondence between name and substance, with an indication of structure as well as chemical composition. However, common (frequently nonsystematic) names remain embedded in nomenclature, often giving rise to duplicate names for the same compound. Although this may be annoying at times, as long as each name unambiguously portrays a specific composition and often a specific structure, it is acceptable. Any treatise on inorganic nomenclature, in order to be useful in the broadest sense, should not only address the problem of naming each substance uniquely, but it should also consider alternative methods that can be used, as well as methods used earlier that are no longer commonly used. These older methods are important for retrospective searching.

## Background

Some years ago the Committee on Nomenclature of the Division of Organic Chemistry of the American Chemical Society (ACS) produced a book on organic nomenclature (*1*). Its chapters were written by various individuals, but the considerable time lag involved in the preparation of the different chapters made it ultimately necessary for an editorial committee to complete and revise the manuscript. Soon after that book was completed, the Nomenclature Committee of the ACS accepted a suggestion that a companion book on the nomenclature of inorganic chemistry would be desirable.

Because the multi-author approach had caused such delays in completing the book on organic nomenclature, W. Conard Fernelius volunteered to prepare a book on inorganic nomenclature. It ultimately became apparent that the preparation of such a treatise was too great an undertaking for one person, and W. H. Powell was recruited as coauthor. Even with this addition, the burden remained onerous, so the third coauthor (B. P. Block) was added. The untimely death of Conard Fernelius, who contributed drafts or outlines for most of the chapters, occurred just as plans were being made to prepare a complete manuscript. His participation was sorely missed in the later stages of the book's preparation. We who completed this book are, of course, entirely responsible for its contents. We trust that our deceased colleague would have given his wholehearted approval to the published version.

In a field such as chemical nomenclature, no single person is responsible for the development of any significant portion of the end product, a set of guidelines for naming chemical species. Although, individually and collectively, we and Conard Fernelius have been active in inorganic chemical nomenclature for many years, we make no pretense that this work is based solely on his or our ideas. We have drawn heavily on the writings of others, either almost verbatim or by paraphrasing, and we trust that our sources are adequately documented. Any omission of credit to others is unintentional.

This work takes as its authorities the Commission on the Nomenclature of Inorganic Chemistry and the Commission on the Nomenclature of Organic Chemistry of the International Union of Pure and Applied Chemistry, the ACS Committee on Nomenclature, and the Chemical Abstracts Service (CAS). In most instances we have avoided taking specific positions with regard to alternative methods available for naming a single compound, usually being content to report viewpoints from different sources. At times the potential that a method may present is noted, even though such usage has not yet developed. In general, when alternative methods of nomenclature are possible, simpler, less informative names are discussed before more complex names are considered. For the most part, alternative names are not given in the examples, and the names presented are intended to exemplify the nomenclature principles under discussion in the section in question.

## Acknowledgments

We are indebted to the Nomenclature Committee of the ACS (K. L. Loening, chairman) for sponsoring this work, to the Nomenclature Committee of the Inorganic Division of the ACS (T. D. Coyle and T. E. Sloan, successive chairmen while the manuscript was being prepared) and other reviewers for valuable comments on the draft manuscript, and to CAS for supporting the preparation of this work in many ways. We express our gratitude in particular to our wives Viola and Sandra and to Naomi Fernelius for their help, encouragement, and patience during the time required to prepare this treatise.

B. Peter Block
East Tennessee State University
Johnson City, TN 37614

Warren H. Powell
Chemical Abstracts Service
Columbus, OH 43210

April 1989

# Chapter 1
# Language and Nomenclature

## 1.1 Introduction

Language has been defined as "the words, their pronunciation, and the methods of combining them, used and understood by a considerable community and established by long usage" or "the vocabulary and phraseology belonging to an art or department of knowledge" (*2*). Language will always be with us. Unless there is some special reason, we use it but think little about it, seeing its faults much more clearly than its strengths. In this respect, the language of any science such as chemistry does not differ from any other language; yet the language of chemistry is a remarkable development. What other science has a vocabulary so systematic? In what other science do the words carry so much meaning? Chemistry is a central science, and its body of knowledge, procedures, and techniques are used widely by other sciences and practical arts. The logic and clear meaning of chemical terms have aided rather than hindered the use of chemistry in other fields.

Nomenclature, the heart of any scientific language, has been defined as "a system or set of names or designations used in a particular science, discipline, or art formally adopted or sanctioned by the usage of its practitioners" (*2*). For present purposes, an appropriate definition of chemical nomenclature is: "a set of chemical names that may be systematic . . . or not and that aims to tell the composition and often the structure of a given compound by naming the elements, groups, radicals, or ions present and employing suffixes denoting function . . . , prefixes denoting composition . . . , configurational prefixes . . . , operational prefixes . . . , arabic numbers or Greek letters for indicating structure . . . , or Roman numerals for indicating oxidation state" (*2*).

Chemical nomenclature should not stand apart from the practice of chemistry. Some are interested only in what to call a compound or substance; some want an official name; others propose names because they do not like recommended names or names in use. A nomenclature suggestion from a prominent chemist is apt to be followed by others, especially those in the same field, even if the suggestion is inconsistent with generally accepted principles of nomenclature.

Officially recommended names are useful only insofar as they are accepted and used by the chemical public. It is very important that there be constant interaction among nomenclature specialists, commissions and committees on nomenclature, and originators of nomenclature proposals on the one hand and those who use the nomenclature on the other, in order that the desires of the latter can be interwoven with general principles of nomenclature and in order that the reasons for certain principles be well understood.

In the absence of such cooperation, the advantages of systematization are soon lost, and ambiguities tend to arise.

## 1.2 Naming Inorganic Substances

Inorganic chemistry has been defined as "a branch of chemistry that deals with chemical elements and their compounds excluding the hydrocarbons and their derivatives but usu. [usually] or often including carbides and other relatively simple carbon compounds . . ." (*2*), and organic chemistry has been defined as "a branch of chemistry that chiefly deals with hydrocarbons and their derivatives" (*2*). Although these two major branches of chemistry developed largely independently, it is no longer possible to distinguish clearly between them because there is a large area of chemistry that can be approached via either discipline. This ambivalence is especially true in nomenclature, even though different basic patterns of nomenclature are usually attributed to each branch.

There have been words to designate inorganic substances from earliest times (*3*), but the work of Lavoisier (1734–1794), generally considered to be the beginning of modern chemistry, brought with it proposals for systematizing nomenclature (*4, 5*) that were published (*6*) by what has been called the "first committee on nomenclature" (*7*). This system was publicized by its use in Lavoisier's treatise (*8*) and further developed by Berzelius (*9*) and Oersted (*10*). It was consistent with the concepts of definite composition, multiple proportions, atoms, molecules, and ions, and it was applicable to a wide variety of types of compounds. Although it failed to reproduce accurately the structure of compounds and was not at all suitable for the rapidly expanding field of organic chemistry, it functioned so well for inorganic substances that it remained dominant for them until near the end of the 19th century. Remnants of this early system still persist today in the names for many oxo acids and their salts.

Because this early system was adequate for a long time, critical examination of the principles underlying inorganic nomenclature did not begin until long after there were serious international efforts to systematize nomenclature for organic compounds. When Werner developed his theory of coordination compounds (*11*), he proposed a system of nomenclature that incorporated Brauner's suggestions (*12*) for characteristic endings to designate the oxidation number of the central element of a coordination compound. This system not only reproduced the stoichiometry of coordination compounds but indicated their structures as well. In a refined form, this system is still the basis for the nomenclature of coordination compounds and is being extended to other types of inorganic compounds insofar as possible.

## 1.3 Nomenclature Principles and Procedures

Uniform nomenclature practices in English language publications were first adopted by the British Chemical Society, which published guidelines for systematic nomenclature in 1882 (*13*). An American Chemical Society (ACS) Committee on Nomenclature and Notation endorsed this system with minor alterations in 1886 (*14*).

The Council of the International Association of Chemical Societies, established in 1911 (*15*), appointed commissions of inorganic and organic nomenclature in 1913 (*16*), but their work stopped during World War I. International nomenclature activities were resumed in 1921 when the International Union of Pure and Applied Chemistry (IUPAC), established in 1919, appointed commissions on the nomenclature of biological, inorganic, and organic chemistry (*17*). The first report of the Commission for the Reform of Inorganic Nomenclature appeared in 1926 (*18*), followed by a comprehensive report in 1940 (*19*). The rules therein were widely adopted and stimulated consideration of further systematization, leading to the publication of the 1957 IUPAC Inorganic Rules (the "Red Book") (*20*). An ACS version with comments was published in 1960 (*21*), and a second edition of the Red Book, *The 1970 IUPAC Definitive Rules for the Nomenclature of Inorganic Chemistry* (*22*), hereinafter called the 1970 IUPAC Inorganic Rules, appeared in 1971, supplemented by a guidebook in 1977 (*23*). Since then several sets of IUPAC recommendations on specific areas of inorganic chemistry have appeared (*24–29*), and a new edition of the Red Book has been prepared (*30*).

Currently, nomenclature activities in the United States are handled by several committees. The ACS Committee on Nomenclature, which reports to and acts on behalf of the Council of the ACS on nomenclature matters, is the coordinating body for all ACS nomenclature activities. It also cooperates in nomenclature matters with the National Academy of Sciences–National Research Council, which is the national adhering organization for IUPAC in the United States. The ACS committee receives recommendations from the nomenclature committees of the various divisions of the Society, reviews the actions of international nomenclature bodies, makes recommendations to international bodies on nomenclature matters, and recommends nomenclature policies for use in the publications of the ACS. Occasionally, although not often, this Committee has chosen not to follow recommendations from international organizations.

Nomenclature principles and procedures arise from a number of sources. The formal procedure for development of chemical nomenclature in the United States is for a proposal to be submitted to an ACS division committee on nomenclature, which forwards the suggestion with any modifications, if needed, to the ACS Committee on Nomenclature. This committee transmits suggestions to appropriate nomenclature commissions of IUPAC. A proposal for inorganic nomenclature is reviewed by the IUPAC Commission on the Nomenclature of Inorganic Chemistry (CNIC) in the context of current methods and developed further and/or modified, as needed. The proposal is then reviewed by the IUPAC Interdivisional Committee on Nomenclature and Symbols to ensure that it is in harmony with nomenclature recommended by other IUPAC divisions. Since 1983 a comprehensive review process (*31*) has been followed to permit widespread consideration of IUPAC recommendations prior to publication.

Neither the ACS nor IUPAC can enforce decisions on nomenclature, although a method for the formation of names, once fully approved, certainly can be considered official. Authors will ultimately render judgment on the nomenclature recommendations of any body, and there are a few instances in which a recommendation by IUPAC has not been generally accepted. Editors of journals are in a position to enforce the use of good, accepted nomenclature; however, they are not uniform in exercising this authority. Some editors are quite rigid in insisting on the use of accepted nomenclature in their publications; others show little concern. In ACS journals the attention given to good nomenclature practice varies considerably from journal to journal.

## 1.4 Nomenclature in *Chemical Abstracts*

In the volume indexes to *Chemical Abstracts* (CA), on the other hand, names must be unique; that is, each name must be the only one that can be generated for a particular substance. This is a stringent requirement that is compounded because decisions on nomenclature matters for the CA indexes cannot wait for national and international considerations. Since 1945 a section on the naming and indexing of chemical substances for the volume indexes to CA has been included either in the introduction to the subject indexes to CA or in the CA *Index Guide* (e.g., *see* ref. 32).

The impression that there are two systems of nomenclature, IUPAC and CA, is not correct. IUPAC recommendations can lead to more than one unambiguous name for a given structure when there is more than one way to generate the name. In such cases, one of the possibilities is usually chosen as the preferred name for use in the CA volume indexes. The Chemical Abstracts Service (CAS) staff works closely with the ACS Committee on Nomenclature and the various IUPAC nomenclature commissions in order to avoid as many differences as possible.

In inorganic nomenclature, one of the most obvious differences is in the spelling of the names for three elements in various parts of the world, that is, aluminum vs. aluminium, cesium vs. caesium, and sulfur vs. sulphur. This treatise follows the practices approved by the ACS Committee on Nomenclature for these names, as well as for the spelling of certain scientific terms, such as stoichiometry as opposed to stoicheiometry.

# Chapter 2
# Fundamental Principles

## 2.1 Introduction

This chapter surveys some general aspects of chemical nomenclature and some basic concepts of inorganic chemistry that directly affect the nomenclature of inorganic substances. It covers three topics: nomenclature conventions, inorganic chemistry, and formulas.

## 2.2 Nomenclature Conventions

Specific ways for naming chemical compounds are usually given in a series of recommendations. Regardless of the pattern by which the name for a compound is formed, it should convey the composition of the compound, identify it unambiguously, and impart information about its structure insofar as possible. Nomenclature is not static, and the standard usage in one period of time may vary somewhat from that in another. There are, however, some general principles, which do not change, that apply to all methods.

Chemical names tend to have a definite morphology, consisting of units that are alike or very similar in form, regardless of the pattern of nomenclature. For each chemical element there is a stem (some elements have more than one; *see* Table A.I, appendix) that may be used in names to indicate the presence of that element. It may be modified by prefixes to indicate the number of atoms present or by suffixes to indicate particular types of substances. In addition, arabic numbers and Roman and Greek letters are used as locants to indicate positions in a structure, and descriptors are used to indicate aspects of structure, such as stereochemical or spatial relationships. Finally, the proper assembly of all these units requires punctuation marks to ensure clarity and exactness of meaning.

### 2.2.1 Types of Chemical Names

*2.2.1.1 Systematic.* This type of name consists of an assemblage of single letters and specially coined or selected groups of letters (usually syllables) known as *morphemes*, each of which has a specific meaning in context (example: penta/sil/ane). Numbers and punctuation marks may also be used as integral parts of a name. Specific patterns provide the composition and often the structure of a compound. A systematic name must be unambiguous, but it does not have to be unique.

*2.2.1.2 Semisystematic.* Only a part of this type of name is used in a systematic sense (*see* Section 2.2.1.1) (example: sodium ferricyanide). The vast majority of chemical names are, strictly speaking, semisystematic rather than systematic. Such names often indicate characteristic groups and can reveal composition or structure or both when additional information about the trivial part of the name is known (example: ferrocene). Semisystematic names for a group of compounds, when based on a trivial name for a parent, can make relationships among a group of compounds evident if the structure of the parent is known (examples: *N,N′*-dimethylborazine and *B,B′*-dimethylborazine). Semisystematic names may give some indication of composition but give no information about structure (example: titanium dioxide), or they may imply structure but give little or no information about composition (example: cubane).

***2.2.1.3 Trivial.*** No part of this type of name is used in a systematic sense (*see* Section 2.2.1.1) (examples: blue vitriol, Zeise's salt, ozone, and lime). Some names of this type are used widely to identify substances, but they do not reveal either composition or structure. The main advantage of trivial names is that they are almost always much shorter than systematic or semisystematic names, whereas the main disadvantage is that each name has to be individually learned. Trademarks, which are used in industry and the marketplace primarily to distinguish one company's products from those of its competitors (example: Hexaphos), are a kind of trivial name that should not be used to designate a particular substance. Likewise, mineralogical names should only be used to designate actual minerals, not to define chemical compositions. Many trivial names are listed in the CA *Index Guide* (*33*), with cross references to the more systematic names used by CAS.

### 2.2.2 Types of Chemical Nomenclature

The generation of names for chemical compounds may be viewed formally as a description of an operation or a series of operations. In some cases a name is simply a summary of constituents. In others it represents the formal addition of an atom or a group of atoms to a compound that has an independent existence, the formal replacement (including substitution) of an atom in a parent compound by another atom or group of atoms, or the formal removal of an atom or atoms from a parent compound. The resulting names may consist of a single word composed of morphemes or two or more words, each of which may be generated by one or more operations.

***2.2.2.1 Additive.*** This type of nomenclature is a method for naming a substance by the formal assembly of names for all of the constituent components of that substance. It is used for binary compounds (example: silicon tetrachloride), pseudobinary compounds (example: phosphoryl trichloride), mixed salts (example: potassium sodium chloride), molecular addition compounds [example: methanol–boron trifluoride(2/1)], coordination compounds (example: hexaamminecobalt trichloride), and functional compounds (example: phosphine oxide) and in a number of other ways in naming organic substances.

***2.2.2.2 Substitutive.*** This type of nomenclature is a method for naming a substance by adding either a prefix or a suffix to the name of a parent compound to indicate formal substitution of an atom or group of atoms for one or more hydrogen atoms of the parent compound. It is the primary method used in naming organic substances and is mainly used in inorganic nomenclature for derivatives of molecular hydrides of nonmetals and other parent compounds (example: trimethylsilanol). It is also used for the organic portions of coordination compounds and other inorganic–organic species [examples: bis(1,2-ethanediamine)copper(II) chloride and hexamethylphosphoric triamide].

*2.2.2.3* ***Replacement.*** This type of nomenclature is a method for naming a substance by combining a prefix or an infix with the name of a parent compound to indicate formal replacement of an atom or group of atoms of the parent compound by another atom or group of atoms. It is used to indicate replacement of skeletal units in rings, chains, and clusters, which can be called skeletal replacement [example: dicarbadodecaborane(12), where the parenthetical number indicates the number of hydrogen atoms in the polyborane], and of =O and/or –OH in oxo acids and their salts, which can be called functional replacement (example: chlorophosphoric acid).

*2.2.2.4* ***Subtractive.*** This type of nomenclature is a method for naming a substance by adding a prefix or suffix to the name of a parent compound to indicate formal removal of an atom, ion, or group of atoms from a parent compound. It is mainly used in inorganic nomenclature for certain boron hydrides [example: 6-debor-*closo*-hexaborane(9)].

*2.2.2.5* ***Functional Class.*** This type of nomenclature is a method for naming a substance by citing a class term as a separate word following the name of a parent compound or a name derived from the latter (example: phosphoric anhydride). When the derived name is the name for a substituent group (formerly called a radical), this method has been called radicofunctional nomenclature. It is used for binary and pseudobinary compounds (example: silyl chloride).

*2.2.2.6* ***Conjunctive.*** This type of nomenclature is a method for naming a substance by joining the names of two molecules; the juxtaposition implies the formal loss of hydrogen atoms from each component. It is used in organic nomenclature for naming compounds in which "a principal group is attached to an acyclic component that is directly attached by a carbon–carbon bond to a cyclic component" (*34a*) (example: 1-naphthaleneacetic acid). It is used in inorganic nomenclature mainly for naming derivatives of polyboranes [example: 1,2-dicarbadecaborane(12)-1-methanol].

*2.2.2.7* ***Multiplicative.*** This type of nomenclature is a method for naming a substance that contains multiple occurrences of identical central atoms or groups of atoms. It is usually an alternative method of nomenclature used for substances with symmetrical structures to produce a simpler name than might otherwise be obtained [example: $\mu_4$-thio-tetrakis(methylmercury)(2+) instead of *Hg,Hg′,Hg″,Hg‴*-tetramethyl-$\mu_4$-thio-tetramercury(2+)].

### 2.2.3 Multiplying Prefixes

Three types of numerical terms are used as multiplying prefixes in chemical nomenclature.

- Prefixes derived from Greek, or occasionally Latin, words for the cardinal numbers are used to indicate the number of identical atoms, unsubstituted substituents, simple ligands, or central atoms in condensed acids, anions of condensed acids, and the skeletons of some molecules and ions.
- Prefixes derived from Greek, or occasionally Latin, words for numerical adverbs are used to indicate a set of identical substituted substituents, ligands, or parent structures or to avoid ambiguity.

- Prefixes derived mainly from Latin words for numbers are used to indicate assemblies of identical ring systems.

These three types will be referred to in this work as "type-1", "type-2", and "type-3" prefixes, respectively. Table A.II lists typical prefixes of these different types. The prefixes "hemi-" for one half, "sesqui-" for three halves, and "sester-" for five halves are used to some extent. The prefixes "ennea-" and "hendeca-" (both derived from Greek number words), which were used earlier for nine and eleven, have largely disappeared (but *see* Section 3.2.2).

There is not complete agreement on when to use type-2 prefixes. Their use is often restricted to expressions containing another prefix {example: bis(dimethylamino)} and to cases in which their absence would cause ambiguity (example: trisdecyl). They are also used more liberally in cases where there is potential ambiguity and to multiply "all complex expressions", as is done in CAS index names.

### 2.2.4 Locants

Locants are symbols specifying the location of structural features that are not implied in the name of the parent compound (examples: substituents; replacement atoms in a chain, ring, or cluster; positions around a center of coordination; unsaturation). They may be arabic numbers, atomic symbols, lower-case Roman letters, Greek letters, or italicized words or morphemes.

*2.2.4.1 Arabic Numbers.* Skeletal positions in a chain, ring, or cluster of atoms are indicated by arabic numbers {often written as "Arabic numbers" and sometimes called Hindu or Hindu–Arabic numbers (*2*)} (example: 2-methyltrisilane).

*2.2.4.2 Atomic Symbols.* Italicized symbols for the elements are used to designate atoms that are not otherwise numbered in a structure and also to designate the atom or atoms through which a ligand is attached to a center of coordination {examples: *N*,*N′*-dimethylsulfuric diamide and bis(1,2,3-propanetriamine-*N*,*N′*)copper(2+)}.

*2.2.4.3 Lower-case Roman Letters.* The 1970 IUPAC Inorganic Rules indicate that the positions of ligands in a coordination polyhedron may be indicated with italicized lower-case Roman letters {example: *abc*,*def*-bis(1,2,3-propanetriamine)cobalt(3+)}. This convention, which was selected because such locants designate positions around a single atom in contrast to the use of arabic numbers to specify the location of a specific atom in a structure (*see* Section 2.2.4.1), is being supplanted by the use of stereochemical descriptors (*see* Section 16.3).

*2.2.4.4 Greek Letters.* Greek letters are used to designate side-chain positions in conjunctive names for organic compounds, positions relative to or distances between functional groups, ring size of lactones, end groups in polymer names, etc. (examples: α-methylcyclohexanemethanol, β-diketone, γ-butyrolactone). Their use to designate positions in a ring or chain has been supplanted by the use of arabic numbers (*see* Section 2.2.4.1) (example: 2-propanol, not β-propanol).

*2.2.4.5 Italicized Morphemes.* In some cases an italicized morpheme is used as a locant to resolve an ambiguous situation (example: [*peroxo*-$^{18}O_2$]peroxodisulfuric acid).

*2.2.4.6 Lowest Locant Set.* The preferred set of locants for a given structural feature (the lower or lowest locant set) is that set of locants with the lower (lowest) locant at the first point of difference after the sets have been arranged in ascending order of priority (usually increasing numerical order). For example, the set 2,1,7,3,9 is preferred to 2,1,6,4,8 because the first difference is that the third term in the ascending sequence 1,2,3,7,9 is lower than the third term in the sequence 1,2,4,6,8. It does not matter that the second set has two locants (6 and 8) lower than the corresponding locants of the first set (7 and 9). If this comparison does not effect a choice, the locant sets are then compared term by term in the order in which they appear in the name. For example, the locant sets 1,3,2,4 and 1,4,2,3 give the same ascending numerical order 1,2,3,4; so 1,3,2,4 is compared with 1,4,2,3. The former has the lower second term (the first difference) and hence is preferred.

### 2.2.5 Descriptor Terms

A number of short words, abbreviations, and symbols are used to indicate some aspect of the structure, often stereochemical, of compounds (*see* Table A.III). These terms are almost always italicized and separated from the rest of the name by a hyphen [examples: *cis*-bis(1,2-ethanediamine)difluorocobalt(1+), *fac*-trichlorotris(pyridine)ruthenium, and *nido*-pentaborane(9)]. In names such as isopropyl and cyclohexane, the descriptors iso- and cyclo- are not italicized and separated by a hyphen because they are considered to be part of the name of the hydrocarbon or its derivatives. In inorganic nomenclature *cyclo*- is normally used as a descriptor term [example: *cyclo*-triphosphate(3–)].

### 2.2.6 Punctuation and Similar Marks

*2.2.6.1 Centered Point.* The centered point is used to separate the components of molecular addition compounds or the component oxides of a mineral in formulas (example: $H_3N{\cdot}BF_3$) and to represent a bond in a structural or semistructural formula (instead of a dash) (example: $H{\cdot}ONO_2$). It is not to be confused with a period.

*2.2.6.2 Colon.* A colon has several uses.

- It is used to separate sets of locants already containing commas, such as locant sets indicating the positions of attachment in bridged polynuclear species (example: 2,3:3,4:4,2-tri-μ-carbonyl-1,1,1,2,2,3,3,4,4-nonacarbonyl-*tetrahedro*-tetracobalt).
- It is used to separate numbers indicating the proportions of constituents in molecular addition compounds [example: methanol compound with boron trifluoride (2:1)] (*see also* Section 2.2.6.7).
- It is used to separate the appended locants designating atoms of bridging groups involved in bonding between polynuclear centers [example: bis-(μ-nonafluorobutyrato-*O*:*O*′)-disilver].

*2.2.6.3 Comma.* The comma is used to separate individual locants, whether numbers, atomic symbols, or Greek letters [examples: bis(1,2,3-propanetriamine-*N*,*N*′)copper-(2+) and 2,3:3,4:4,2-tri-μ-carbonyl-1,1,1,2,2,3,3,4,4-nonacarbonyl-*tetrahedro*-tetracobalt]. Commas are not used to separate individual letter locants indicating positions around a central atom, although they are used to separate series of such letter locants {examples: *ab*-dichloro-*defc*-{[2-(methylthio)ethyl]iminodiacetato-*O,O′,N,S*}platinum and *abc,def*-bis(1,2,3-propanetriamine)cobalt(3+)}.

*2.2.6.4* ***Dash.*** A dash (rule, long hyphen), which is longer than the hyphen (*see* Section 2.2.6.5), has several uses.

- It is used to indicate a bond in structural formulas (example: $H_2N–BF_2$).
- It is used to separate the two components in the name of a molecular addition compound [example: ammonia–boron trifluoride(1/1)].
- It is used between italicized atomic symbols to indicate metal–metal bonds [example: bis(pentacarbonylmanganese) (*Mn–Mn*)].
- With η, the dash is used between two number locants to indicate the inclusive atoms involved in linkages to a coordination center [example: (1–3-η-but-2-enyl)tricarbonylcobalt].

*2.2.6.5* ***Hyphen.*** The hyphen, which is not to be confused with the dash (*see* Section 2.2.6.4), also has several uses.

- It is used to set off locants, locant sets, and descriptors from the rest of the name (example: 1-chlorotriazene).
- It is used to separate bridging groups from the rest of the name (IUPAC) [example: tri-μ-carbonyl-bis(tricarbonyliron)].
- It is used to separate the two different kinds of locants in the names of some fused-ring systems (example: pyrido[2,3-*b*]-1,6-naphthyridine).

Using a hyphen between vowels (example: penta-ammine instead of pentaammine), as a substitute for enclosing marks [example: bis-1,2-ethanediamine-copper(2+) ion instead of bis(1,2-ethanediamine)copper(2+)], or for separating prefixes [example: bromodichloro-methylsilane instead of bromodichloro(methyl)silane] is not considered good practice in the United States.

*2.2.6.6* ***Semicolon.*** The semicolon is used to separate sets of locants designating ligating atoms of chelating ligands in a polynuclear coordination entity {example: bis(μ-diphenylphosphinato-1κ*O*:2κ*O′*)-tetrakis(2,4-pentanedionato)-1κ$^2$*O,O′*;1κ$^2$*O,O′*;-2κ$^2$*O,O′*;2κ$^2$*O,O′*-dichromium for $(AcCHAc)_2Cr[OP(C_6H_5)_2O]_2Cr(AcCHAc)_2$}. Note that a colon cannot be used because it would indicate a bridging group.

The semicolon is also used to separate sets of locants containing colons [example: *cyclo*-hexachloro-1κ$^2$*Cl*,2κ$^2$*Cl*,3κ$^2$*Cl*-tri-μ-nitrido-1:2κ*N*;1:3κ*N*;2:3κ*N*-triphosphorus for $(Cl_2PN)_3$].

*2.2.6.7* ***Slash.*** The slash (solidus, diagonal) is used to separate the number of molecules of each type in a molecular addition compound [example: methanol–boron trifluoride-(2/1)] (*see also* Section 2.2.6.2). It has also been used in the literature to separate the names of the individual constituents of mixtures (example: ethanol/water).

*2.2.6.8* ***Space.*** Spaces are used to separate the words in names of compounds consisting of two or more distinct words, such as names for binary compounds and names that include the functional class (examples: boron trifluoride and sulfuric acid). These spaces are retained when such species are named as ligands in coordination compounds [example: *trans*-bis(dipropyl sulfide)dinitroplatinum].

## 2.2.7 Enclosing Marks

Three sets of enclosing marks are used in chemical nomenclature: parentheses ( ), brackets [ ], and braces { } (curved, square, and curly brackets, respectively, in British terminology). There are two attitudes toward the use of enclosing marks.

One attitude is to use them only when necessary for clarity, whereas the other is to use them every time a type-2 numerical prefix (*see* Table A.II) is used. The former, for example, would lead to omission of enclosing marks in such cases as trisdecyl and tetrakishydroxymethyl.

Parentheses are used if only one kind of enclosing mark is needed. The need for more than one set of enclosing marks within another necessitates an order of precedence or nesting order. The normal order is parentheses within brackets and brackets within braces; that is, {[( )]}. Braces, however, are not used in CAS index nomenclature. Brackets and parentheses have specific meanings in names for isotopically modified compounds (*26*), and brackets are used in organic ring nomenclature in ways that have no parallel in inorganic nomenclature.

For clarity or emphasis, enclosing marks may be used in formulas to enclose sets of identical groups of atoms {examples: $Al_2(SO_4)_3$ and $P[N(CH_3)_3]_3$}, with the normal nesting order when several are needed. An exception to the nesting order is made in formulas for coordination compounds, for which the convention established by Werner to enclose the formula of a neutral or ionic coordination entity in brackets is followed. Hence, in formulas for coordination compounds the nesting order for enclosing marks is [(· · ·)], [{(· · ·)}], [{[(· · ·)]}], [{{[(· · ·)]}}], etc.

## 2.2.8 Greek Letters

Greek letters are used in chemical nomenclature for many purposes in addition to their use as locants already described in Section 2.2.4.4.

***2.2.8.1 Allotropes.*** Allotropic or polymeric forms of substances have been distinguished by Greek letters (examples: α- and β-quartz, λ- and μ-sulfur).

***2.2.8.2 Alloys.*** Alloy phases have been designated by Greek letters (example: α- and β-bronze).

***2.2.8.3 Other Uses.*** Specific meanings are attached to a number of Greek letters.

▶ Delta:

- δ indicates conformation of chelate rings (paired with λ).
- Δ indicates chirality of certain chelates (paired with Λ) {example: The isomer of $[Co(en)_3]Cl_3$ that gives positive rotation at 589 nm is designated Λ δδδ}.
- Δ has also been used to indicate the presence of a double bond (1) in a heteromonocyclic ring (example: $\Delta^3$-1,2-diazetine), (2) between two rings or ring systems

in a ring assembly name (example: $\Delta^{2,2'}$-bipiperidine), or (3) between a ring or a ring system and an acyclic component in a conjunctive name (example: $\Delta^{2,\alpha}$-pyrrolidineacetic acid).

- $\delta^c$ indicates that "c" double bonds are attached to a skeletal atom of a cyclic parent hydride (usually found together with $\lambda^n$, q.v.) [example: 3-oxo-1$\lambda^4\delta^2$,3$\lambda^4$,4,2,5-trithiadiazole for

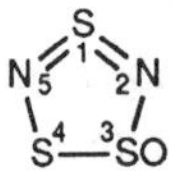

▶ Eta:

- $\eta^n$ indicates the number of atoms of an organic group that are bonded to a center of coordination [example: bis($\eta^5$-cyclopentadienyl)iron].

▶ Kappa:

- κ followed by an atomic symbol designates a specific atom or atoms through which a ligand is linked to a center of coordination [example: *cis*-bis(glycinato-$\kappa^2 N,O$)palladium].

▶ Lambda:

- λ indicates conformation of chelate rings (paired with δ).
- Λ indicates chirality of certain chelates (paired with Δ).
- $\lambda^n$ indicates that a skeletal atom of a parent hydride has the nonstandard bonding number "*n*" (example: $\lambda^4$-sulfane for $SH_4$).

▶ Mu:

- μ indicates a bridging group [example: tri-μ-iodo-bis(tricarbonylrhenium)].

▶ Pi:

- π designates a bond arising from the overlap of p orbitals.

▶ Sigma:

- σ designates a bond arising from a shared pair of electrons.

### 2.2.9 Ordering Principles

***2.2.9.1 Electronegativity.*** Since early in the 19th century, chemists have recognized an intimate relationship between chemistry and electricity. They noted in particular that the atoms of some elements form anions, whereas those of others form cations. This relationship is important in nomenclature because in Germanic languages the cation is traditionally cited before the anion (examples: sodium chloride, calcium sulfate). This same principle is used in forming names for nonelectrolytes; that is, the more electropositive element is cited first, followed by the more electronegative element (examples: carbon dioxide and boron trifluoride). Thus, for nomenclature purposes, a relative order of electronegativity is needed.

Unfortunately, there are minor variations in scales of electronegativities derived by various methods, so an arbitrary order has been adopted (*22a, 22b*). The order of elements given by the continuous line through the periodic table shown in Figure A.1 is generally consistent with the various scales of electronegativity, is easily visualized by its relation to the long form of the periodic table, and is consistent with the rules for naming binary compounds.

Two exceptions to a continuous line that moves down one family after another (that is, hydrogen and oxygen) make this continuous line conform quite closely to most experimentally determined electronegativity orders. The Group 18 elements (*see* Section 3.2.6 for a discussion of the designation of the groups in the periodic table) have not been included in Figure A.1. They clearly are neither more electropositive than the Group 1 elements nor more electronegative than the Group 17 elements, so they are neither an appropriate extreme right nor extreme left column. The Group 18 elements were placed to the left of the Group 1 elements in the general element sequence table included with the 1970 IUPAC Inorganic Rules (*22c*). Although the continuous line in that table leads to the conclusion that Group 18 elements are the most electropositive elements, a caveat in the text nullifies such a conclusion (*22b*).

***2.2.9.2 Alphabetical Order.*** This common ordering method is used in ordering cations, anions, substituents, ligands, prefixes, infixes, etc. The name of each segment is considered to begin with the first letter of the name of that segment, and any multiplicative prefixes not an integral part of the name of the segment are ignored. Because "bis(sulfide)" refers to two sulfide ions, it would be alphabetized under "s", whereas the disulfide ion $S_2^{2-}$ would be alphabetized under "d". Similarly, diammine, referring to two ammonia ligands, would be alphabetized under "a", whereas dimethylamine would be alphabetized under "d".

***2.2.9.3 Cahn–Ingold–Prelog Sequence Rule.*** The Cahn–Ingold–Prelog (CIP) sequence rule (*35*) is used to determine the priority ranking of atoms or groups attached to a chiral element in assigning the absolute terms *R* and *S* to the stereoisomers. This ranking depends first on the descending order of atomic number of the atoms directly attached to the chiral center. Other criteria are invoked as necessary in order to make an unequivocal assignment. Although the CIP sequence rule was originally developed for use with organic compounds, it also has been used in a system for naming inorganic stereoisomers.

## 2.3 Concepts in Inorganic Chemistry

### 2.3.1 Types of Inorganic Compounds

For the purposes of nomenclature, inorganic compounds may be divided into the following types.

***2.3.1.1 Binary.*** These compounds are substances composed of two different elements (examples: NaCl and $K_2S$).

***2.3.1.2 Pseudobinary.*** These compounds are substances composed of more than two different elements that can be treated as if they were binary (examples: $Li_2SO_4$, $NH_4F$, and NaCN).

***2.3.1.3 Functional.*** These compounds are substances containing an atom or group of atoms that reacts in a characteristic way [examples: $H_2SeO_4$, CsOH, and $H_2NC(O)NH_2$].

***2.3.1.4 Parent.*** These compounds are substances with at least one substitutable hydrogen atom attached to a skeletal atom or characteristic group (*see* Section 2.3.1.3) or at least one characteristic group capable of forming derivatives [examples: $H_2NC(O)NH_2$ and $H_2P(O)OH$]. (For this purpose an acidic hydrogen atom is not usually considered to be a substitutable hydrogen atom.)

*2.3.1.5 **Parent Hydride.*** These compounds are a subset of parent compounds (*see* Section 2.3.1.4) consisting of a central atom or a polynuclear system of skeletal atoms to which only hydrogen atoms are attached (examples: $SiH_4$, $N_2H_4$, and $B_{10}H_{14}$).

*2.3.1.6 **Coordination.*** These compounds are substances containing entities consisting of one or more central atoms surrounded by other atoms or groups of atoms called ligands, at least one of which is bonded to the central atom by a coordinate covalent bond {examples: $K_2[PtCl_6]$, $[Cr(NH_3)_6]Cl_3$, and $[Co(NH_3)_3Cl_3]$}.

### 2.3.2 Oxidation Number

Since the establishment of methods for determining formulas of compounds, chemists have been impressed by the constancy of the combining ability of the atoms of an element in a series of compounds. They have at various times used the terms *wertigkeit*, valence, valency, oxidation state, and oxidation number to designate this property. Although each of these names has acquired undesirable connotations and lost popularity, oxidation number seems to be the least objectionable. The *oxidation number* of the atoms of an element in a binary compound (and, by extension, in ternary and higher compounds) is a theoretical concept indicating the charge that would be present on each atom of the element if the electrons in each bond involving that element were assigned to the more electronegative partner. Hydrogen in combination with nonmetals is considered to be positive. Atoms in polyatomic forms of elements are considered to have an oxidation number of zero, and a bond between structurally equivalent atoms of the same element makes no contribution to the oxidation number.

Oxidation numbers, when expressed as Roman numerals (both positive and negative) or as zero and attached in parentheses to the element's name or symbol, are called Stock numbers (*36*) (*see* Table 2.I).

Although the concept of oxidation number is deeply rooted in inorganic chemistry, the concept is purely theoretical and sometimes, if followed too rigorously, leads to difficulty. Some cases can be resolved by recognizing that the same element may have different oxidation numbers in the same compound. For instance, the apparent oxidation number of II for sulfur in $Na_2S_2O_3$ is not correct because the two sulfur atoms are not equivalent. The more informative formula $Na_2[SSO_3]$ can be interpreted to indicate that there is a central sulfur atom surrounded by three oxygen atoms and one sulfur atom that is equivalent to the oxygen atoms. Hence, the central sulfur atom has an oxidation number of VI if the sulfur atom equivalent to the oxygen atoms is considered to be –II in analogy to the oxygen atoms. There is no such simple solution for the series of ions $[Ni(Ph_2C_2S_2)_2]^{n-}$, in which $n$ may have the values 2, 1, and 0.

**Table 2.I. Oxidation Numbers**

| *Elements in* | *when considered as* | *have oxidation numbers of* |
|---|---|---|
| HCl | $H^+$ and $Cl^-$ | H = I ; Cl = –I |
| $H_2O$ | $2H^+$ and $O^{2-}$ | H = I ; O = –II |
| $CO_2$ | $C^{4+}$ and $2O^{2-}$ | C = IV ; O = –II |
| Hg | Hg | Hg = 0 |
| $S_8$ | 8S | S = 0 |
| FOOF | $2O^+$ and $2F^-$ | O = I ; F = –I |
| $NH_4^+$ | $4H^+$ and $N^{3-}$ | H = I ; N = –III |
| $PdBr_2(NH_3)_2$ | $Pd^{2+}$, $2Br^-$, $6H^+$, and $2N^{3-}$ | Pd = II ; Br = –I<br>H = I ; N = –III |

### 2.3.3 Ewens–Bassett Number

An alternative to the Stock number (*see* Section 2.3.2) for indicating the proportions of the constituents of a compound is the Ewens–Bassett number (*37*). It was designed for ionic compounds and is simply the amount of charge on an ion, followed by the sign of the charge. It has the advantage of being an experimentally determinable number, whereas oxidation number is a theoretical concept. Zero is not used in the Ewens–Bassett system because it is used as a Stock number. Some examples are shown in Table 2.II.

**Table 2.II. Ewens–Bassett Numbers**

| *Species* | *Ewens–Bassett Number* |
|---|---|
| $Fe^{3+}$ | 3+ |
| $[Co(NH_3)_6]^{3+}$ | 3+ |
| $SO_4^{2-}$ | 2− |
| $[Ni(CN)_4]^{4-}$ | 4− |

### 2.3.4 Terms Used in Coordination Chemistry

***2.3.4.1 Coordination Number.*** The number of atoms directly linked to a central atom in a coordination entity is called the coordination number (i.e., the number of closest neighbors). A closest neighbor may be a single atom (ion) or the closest atom of a group. Such numbers have general significance and are used for a variety of purposes. Occasionally, especially in organometallic compounds, the use of the term coordination number is not entirely consistent with the classical usage in coordination compounds because two or more atoms of an unsaturated ligand may be considered to occupy a single coordination position.

In crystallography, coordination number indicates the number of neighbors nearest to a given atom or ion. As long as there are definite bonds between an atom or ion and all of its nearest neighbors, the two usages are consistent. However, for units held together by purely electrostatic attraction, the usages can be different. For example, in solvates and complex compounds, the sodium ion never has a coordination number greater than 4, although the crystallographic coordination number of sodium in a crystal of NaCl is 6.

***2.3.4.2 Other Terms.*** Some terms other than "coordination number" that are used in connection with coordination entities are defined as follows.

- Complex: the term once used almost exclusively to indicate a coordination entity (no longer considered desirable).
- Coordination Center: central atom in a coordination entity.
- Ligancy: another term for coordination number.
- Ligand: atom or group of atoms attached to a coordination center by any type of bond.
- Ligating Atom: atom through which a ligand is attached to a central atom.
- Donor Atom: another term for ligating atom.
- Multiplicity of a Coordination Center: the number of coordination centers in a coordination compound.
    - Mononuclear: term indicating one central atom.

   - Polynuclear: term indicating more than one central atom. Dinuclear indicates two central atoms, trinuclear three, tetranuclear four, etc.
- Multiplicity of a Ligand: the number of ligating atoms in a ligand.
   - Monodentate: term indicating that a ligand is attached to a central atom through one ligating atom.
   - Multidentate (sometimes polydentate): term indicating that a ligand is attached to a central atom through two or more ligating atoms. Bidentate indicates two, terdentate (sometimes tridentate) three, quadridentate (sometimes tetradentate) four, etc.
- Denticity of a Ligand: another term for multiplicity of a ligand.
- Chelating Ligand: a multidentate ligand attached to a central atom through more than one ligating atom.
- Chelate: a coordination entity containing at least one chelating ligand.
- Chelate Ring: cyclic group of atoms formed by a chelating ligand and a central atom.
- Flexidentate: a multidentate ligand capable of attaching to a central atom through different sets of ligating atoms.
- Ambidentate: a monodentate group capable of linking to a central atom through different ligating atoms.

## 2.4 Formulas for Heteroatomic Species

### 2.4.1 General

The information necessary for naming a chemical species is summarized in the *chemical formula*, which will hereafter be referred to as the "formula", for that species. The formula is a collection of element symbols (*see* Section 3.2.4) with right subscripts giving the relative numbers of atoms of different kinds of elements in the entity in question. When the molecular weight of a neutral species is unknown or cannot be determined, it is possible to write a formula that gives only the relative numbers of atoms in that species. Such a formula is known as an *empirical formula* (example: NaCl). When the molecular weight is known and the formula gives the actual numbers of atoms in a molecule of the species in question, the formula is a *molecular formula* (example: $N_2O_4$).

*Structural formulas* are used in order to represent the spatial arrangement of atoms in a molecule.

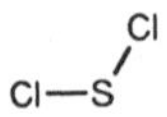

Cl—S—Cl (S bonded to two Cl; second Cl drawn angled upward)

When complete structural formulas are not needed to show a detail of structure or when the space required for the structural formula is too great, *semistructural formulas* {example: line formulas such as $[Co(NH_3)_3(NO_2)_3]$} are used. Italicized prefixes that give structural information (*see* Table A.III) may be connected with the formula by a hyphen {example: *trans*-$[Pt(NH_3)_2Cl_2]$}. Formulas for ionic species may similarly impart information ranging from empirical relationships to detailed structural relationships.

It is customary to use abbreviations for many common groups of atoms [examples: Ac for $CH_3CO$, Bu for $C_4H_9$, Bz for $C_6H_5CO$ (not $C_6H_5CH_2$), Et for $C_2H_5$, Me for $CH_3$, Ph for $C_6H_5$, and Pr for $C_3H_7$]. Although some of these abbreviations are the same as some element symbols, the context usually prevents confusion.

### 2.4.2 Order of Symbols

There are conventions for the ordering of symbols in a formula, just as there are for the ordering of the components in a name. The main principle is that the electropositive constituent (the cation) should always be placed first. In binary entities consisting of a metal and a nonmetal, the metallic element is always placed first. The established order for binary compounds in which both elements are nonmetals is to place first that element appearing first in the sequence Rn, Xe, Kr, B, Si, C, Sb, As, P, N, H, Te, Se, S, At, I, Br, Cl, O, F. With the exception of the Group 18 gases, this is the increasing order of electronegativity given in Figure A.1.

The order in species containing three or more elements depends upon what is known about the species. In salts, cation symbols are listed in alphabetical order of symbols, followed by a separate list of anion symbols, also in alphabetical order of symbols (example: $KMgClF_2$). Acids are treated as hydrogen salts; that is, hydrogen is a cation, but hydrogen is listed last in a sequence of cations (example: $KHF_2$). If two or more different atoms or groups are attached to a single central atom, the symbol of the central atom is placed first, followed by the symbols of the remaining atoms or groups in alphabetical order [example: $P(NCO)_3O$]. If three or more atoms form a chain, the sequence should be in accordance with the order in which the atoms are actually bound (example: HSCN). Symbols for atoms in groups that have structural identity are placed together and arranged as appropriate for a group with a central-atom structure or a chain structure (examples: $SO_4{}^{2-}$ and $SSS^{2-}$). The constituents in intermetallic compounds are placed in alphabetical order of their symbols (example: $Cu_4Hg_2Sn$). Loosely bound molecules are cited in increasing order of multiplicity, which is indicated by citing arabic numbers before their formulas (example: $8CHCl_3 \cdot 16H_2S \cdot 136H_2O$).

Many formula indexes use an arbitrary order of symbols in formulas, so it is necessary to find out what convention is used in any given index. The Hill system (that is, putting C first, followed by H, and then listing the symbols of any remaining elements in alphabetical order) is used in the CA *Formula Index*.

### 2.4.3 Structural Formulas

The structure of covalent compounds has long been shown by using a line (dash) between atomic symbols to represent a single bond or a shared pair of electrons (*see also* Section 2.2.6.4) and a double or triple line to represent a double or triple bond. Although it is recognized that the sharing of multiple pairs of electrons cannot represent the complexities of bonding, structural nomenclature is still based on the concept of single and multiple bonds. That the latter are really more or less complex must be understood. Tautomerism, resonance, and the existence of fluxional molecules all produce nomenclature problems insofar as a single systematic structural name is concerned. The individual forms can be named, but there is as yet no way except through trivial names to indicate that phenomena such as those listed are present.

Representing electron-deficient cluster compounds and some coordination compounds by structural formulas is also troublesome, and it has become the custom to represent such structures with lines between adjacent atoms. In this manner, geometrical forms are easily represented; however, the number of bonding electrons actually present is not equal to the number of lines. Nomenclature for such species generally does not present the same kind of problems as does that for species with multiple forms.

# Chapter 3
# Homoatomic Species

## 3.1 Introduction

The chemical elements (i.e., the fundamental substances, each of which consists of atoms all having the same atomic number) may be monoatomic or polyatomic, and a given element may exist in more than one polyatomic (molecular) form. The different polyatomic forms of an element (i.e., its allotropic modifications) have different names, either trivial or systematic. This chapter will be devoted to the nomenclature of *homoatomic species*, that is, neutral and ionic monoatomic and polyatomic species derived from one element.

The name for an element is a general term for a substance; it carries no implications about the atomic arrangement of the substance (structure). Thus it is necessary to speak of an atom, molecule, cation, or anion of an element (examples: a hydrogen atom, a hydrogen molecule, a hydrogen cation, or a hydride anion). Names for homoatomic ions, radicals, and molecules are usually derived from the element names, but occasionally from a Latin equivalent.

## 3.2 Names and Symbols for the Elements

### 3.2.1 Traditional Names

The right of the discoverer to name the object discovered is not questioned in many scientific fields and is limited only slightly in others. For elements this right has been largely retained, and the naming of most elements has proceeded smoothly. The elements that were known in antiquity bear names of ancient origin, whereas most of the rest of the elements are known by names suggested by the discoverers. During the period 1911–1947, however, there were multiple claims to the discovery of elements with atomic numbers 43, 61, 72, 85, and 87, resulting in some controversy over the right to name them.

CNIC was given the task of resolving those cases that had not been settled by the chemical community and was also asked to recommend single names for common elements that had different names in different languages. Although it was successful with the first task, CNIC had little success with the second, even for scientific purposes. A group of German chemists charged with translating the 1970 IUPAC Inorganic Rules into German, however, took a very significant step by proposing the acceptance of Hydrogen, Oxygen, Nitrogen, and Carbon for Wasserstoff, Sauerstoff, Stickstoff, and Kohlenstoff and the alteration of the spelling of Jod to Iod and Wismut to Bismut (*38,*

*39*). CNIC remains the IUPAC commission charged with the task of recommending approved names for elements. Its decisions, according to the 1970 IUPAC Inorganic Rules (*22d*), are "based upon considerations of prevailing usage and practicability" and "their selection carries no implications regarding priority of discovery."

The procedure by which the elements have been named has led to a collection of names that are not systematically related to one another, so it is possible only to compile lists of approved names, not rules for constructing approved names. Such a list for all recognized elements, with ACS-approved spelling of the names, is given in Table A.I in the appendix. Some alternative names are included because they are likely to be encountered in the literature or because they are the basis for the symbol for the element or for a stem used for naming compounds of the element.

The only generality to be made about the names for the elements is that the ending -ium (-um in three instances) usually indicates a metal. Helium, selenium, and tellurium are exceptions, and a number of metals (*see* Section 3.2.6) do not have names ending in -ium or -um, even in alternative forms (examples: bismuth, nickel, and zinc).

### 3.2.2 Systematic Names

The first nine transuranic elements were named without serious problems according to the right-of-discovery principle. A potential conflict involving element 102 was avoided when the real discoverers accepted the name nobelium, which had been proposed by the group that had erroneously reported its discovery.

Significant controversy, however, has arisen with regard to the discovery (first isolation) of elements with atomic numbers 104 and 105. In 1974 IUPAC, in collaboration with the International Union of Pure and Applied Physics (IUPAP), agreed to appoint a committee of experts that would include three members each from the United States and the Soviet Union and three (including the chairperson) from other countries to try to resolve the question of priority of discovery of these elements. Although to date this committee has not issued a report, the three U.S. members of the committee have published their views (*40*), which include published criteria for the discovery of new elements (*41, 42*). However, the work of another group, an ad hoc joint IUPAC–IUPAP working group on the transfermium elements, has been reported (*43*). It is interesting that reports of the discovery of elements with atomic numbers of 106 and higher have appeared, but that no names have been proposed for them.

While the controversy over the naming of elements 104 and 105 continues, a considerable volume of literature concerning these elements has accumulated, and the number of reports concerning elements with even higher atomic number has increased. Hence, in 1979 CNIC published a systematic method for naming elements (*25*) to make it possible to refer to an element before there was either experimental evidence for its existence or agreement on its name. This system is based on a mixture of Greek and Latin numerical terms for the individual digits of the atomic number. To provide unique symbols for the elements involved (*see* Section 3.2.4), the system is designed so that the 10 terms have different initial letters. These numerical terms are

| | | | | | | | | | |
|---|---|---|---|---|---|---|---|---|---|
| 1 | un | 3 | tri | 5 | pent | 7 | sept | 9 | enn |
| 2 | bi | 4 | quad | 6 | hex | 8 | oct | 0 | nil |

The names are formed by citing these terms in the order of the digits that make up the atomic number, followed by the ending "ium". The final "i" of "bi" and "tri" is elided when it occurs before "ium", as is the final "n" of "enn" before "nil". For example, element 107 is named unnilseptium (un-nil-sept-ium), element 190 unennilium (un-en-nil-ium), and element 263 bihextrium (bi-hex-tr-ium). In these names, each term is to be pronounced separately. These systematic names have not yet been widely accepted, in spite of statements that each name is to be used only until such time as the chemical community agrees on a trivial name for the element in question. In general, physicists have totally rejected the system. The use of atomic numbers instead of names, which is another possibility, makes the naming of all but the simplest compounds rather cumbersome.

### 3.2.3 Stems

Stems characteristic of each element are used in the construction of names for inorganic compounds. These stems (there are more than one for some elements) are indicated in Table A.I by hyphens and slashes. Unfortunately, there is no system for choosing which stem to use when there are multiple stems. Secondary stems were usually introduced to produce distinctive names when two different nomenclature systems produced the same name for different compounds.

### 3.2.4 Symbols

For many purposes the traditional or alternative names of elements are not used, but instead symbols are used to represent the elements. These symbols, capital Roman letters often followed by lower-case Roman letters, are usually derived from the names of the elements. For elements with atomic numbers greater than 103, three-letter symbols are derived from the systematic names by citing the initial letter of each of the numerical terms in the systematic name (examples: Uns for unnilseptium, Uen for unennilium, and Bht for bihextrium) (*see* Section 3.2.2). The symbols for the elements, which, in contrast to the names, are truly international, are also listed in Table A.I. In all but one case, when the name and symbol for an element are not related, the symbol is derived from the Latin name for that element. The Latin names involved are included in Table A.I. The exception is the symbol W for tungsten, which is derived from that element's alternate name, wolfram, a name that is preferred in Germanic languages.

Subscript and superscript characters used with the symbol of an element give the following specific information.

- mass number—left superscript
- ionic charge—right superscript
- atomic number—left subscript
- number of atoms of element—right subscript

For example, the symbol ${}^{200}_{80}Hg_2{}^{2+}$ represents a species bearing an ionic charge of 2+ and consisting of two atoms of mercury, each with atomic number 80 and mass number 200.

Normally only the right subscript denoting the number of atoms of the element is used in chemical formulas (*see* Section 2.4) (example: $Na_3PO_4$). When no subscript number is given, it is assumed that only one atom is present. Right superscripts are also used when it is desired to indicate the presence of ions [example: $(Na^+)_3(PO_4)^{3-}$] or to give the formula for an ion (example: $SO_4^{2-}$). The right superscript is written $n+$ or $n-$; that is, the number representing the quantity is given before the sign of the charge, not after it. The atomic number is, of course, obvious from the atomic symbol, so the use of the left subscript to indicate atomic number is redundant and not needed for most purposes.

### 3.2.5 Isotopes

Individual isotopes of an element may be named by attaching the appropriate mass numbers to the name of the element with a hyphen (*22e*) and may be represented by the atomic symbol of the element with the mass number as a left superscript (*see* Section 3.2.4). Thus, the major isotopes of chlorine are named chlorine-35 and chlorine-37 and have the symbols $^{35}Cl$ and $^{37}Cl$. Early investigators used separate names for each isotopic nuclide, but trivial names have survived only for the isotopes of hydrogen (i.e., protium, deuterium, and tritium rather than hydrogen-1, hydrogen-2, and hydrogen-3 for $^{1}H$, $^{2}H$, and $^{3}H$, respectively). The symbols D and T have been used for deuterium and tritium, but $^{2}H$ and $^{3}H$ are preferred in formulas because D and T disturb alphabetical ordering.

### 3.2.6 Classification of Elements

Most classifications of the elements are based on chemical- or electronic-structure similarities. These classifications can generally be related to periodic tables, one form of which is reproduced in Figure A.2. The numbering of the vertical groups of elements in periodic tables from Group I to Group VIII is well established internationally, but the use of A and B to designate the groups and subgroups has not been consistent (*44*). Although the 1970 IUPAC Inorganic Rules do not contain a periodic table, they do state that the subgroups starting with K, Ca, Sc, Ti, V, Cr, and Mn should be designated 1A, 2A, etc., and those starting with Cu, Zn, Ga, Ge, As, Se, and Br should be designated 1B, 2B, etc. (*22f*). The ACS Committee on Nomenclature recommended that the use of A and B to designate groups and subgroups of periodic tables be abandoned and that these groups be numbered from 1 to 18, as indicated in Figure A.2 (*45*). (The use of d and f recommended for Groups 3 through 12 in ref. 45 has been omitted.) There has been considerable criticism of this recommendation, and whether it will ultimately be accepted or rejected by the chemical community is uncertain at this time.

CNIC included the 1–18 numbering for groups and subgroups of periodic tables in the new edition of the IUPAC Rules for Inorganic Nomenclature, but it has also included alternatives (*30*). The 1–18 numbering scheme will be used in this treatise, although the two widely used A–B schemes are also indicated in Figure A.2.

The oldest classification of elements is as metals and nonmetals, a differentiation denoted by the stepped line shown in the periodic table in Figure A.2. The distinction between these two categories is not a sharp one, and some of the elements next to the line are often called semimetals. CNIC recommends against the use of "metalloid" for these elements because it is not used in the same sense from language to language (*22g*).

The broad classification of elements into main, transition, and inner transition elements is widely used. This classification is usually based on position in periodic tables and is related to similarities in electronic structure. In the discussion that follows, some major electron shells may be considered "complete" in one series, even though electrons are later added to d or f orbitals of that shell in another series of elements. The main-group elements, which have also been called characteristic, typical, or representative elements, are the elements in Groups 1, 2, 13, 14, 15, 16, 17, and 18. All of the electrons responsible for the chemical behavior and for most of the physical properties of these elements are in orbitals having the major quantum number $n$ (s and p orbitals) (Bohr classification: outer shell incomplete).

The term "transition element", originally used to designate the elements in Groups 8, 9, and 10 for which there is no parallel among the main-group elements, is now widely used to denote those elements in Groups 3 through 12. Transition elements are, with the exception of the Group 12 elements, those elements in which the chemically active electrons are in orbitals with major quantum numbers $n$ (s and p orbitals) and $n - 1$ (d orbitals) (Bohr classification: two outermost shells incomplete except for the elements in Group 12). Because zinc, cadmium, and mercury have some properties similar to those of the preceding transition metals in the same series, they are usually included among the transition elements, although they each have complete d orbitals in the $n - 1$ shell and do not form monoatomic cations with more than one oxidation number. The inner transition elements (other than lutetium and lawrencium) each have an incomplete $n - 2$ major shell (f orbitals) in addition to incomplete $n$ and $n - 1$ shells (Bohr classification: three outermost orbitals incomplete except for lutetium and lawrencium).

A more recently introduced terminology, for which there is no uncertainty in terms and no dual usage, is

- s-block elements: Li, Be, and their congeners
- p-block elements: B to Ne and their congeners
- d-block elements: Sc to Zn and their congeners
- f-block elements: Ce to Lu and their congeners

The following are collective names for some groups of elements in periodic tables:

- Alkali metals: Li, Na, K, Rb, Cs, Fr
- Alkaline earth metals: Ca, Sr, Ba, Ra
- Rare earth elements: Sc, Y, La through Lu
- Lanthanoid or lanthanide elements: Ce through Lu
- Actinoid or actinide elements: Th through Lr
- Chalcogens: O, S, Se, Te, Po (chalcogenides in binary compounds)
- Halogens: F, Cl, Br, I, At (halides in binary compounds)
- Noble, inert, or rare gases: He, Ne, Ar, Kr, Xe, Rn

Many now consider the alkaline earth metals to include Be and Mg in addition to Ca through Ra (i.e., to include all the elements of group 2) (*46*). This group originally consisted only of the elements Ca through Ra (*22f*).

CNIC recommended the use of the terms "lanthanoid" and "actinoid" (*22f*) to circumvent objections to the use of the words "lanthanide" and "actinide" because the suffix "-ide" has a specific meaning in the naming of monoatomic anions (*see* Section 3.4.2), whereas "-oid" is a suffix meaning resembling or having the quality of (*2*). However, these terms have not been widely used. No one of the names for the helium family of elements is used in preference to the others, which suggests that none is really appropriate.

There have been attempts to coin other group names, but only the term pnicogens (pnictides in binary compounds) for N, P, As, Sb, and Bi has received much use. Its use is not recommended by CNIC, and, in order to forestall the introduction of other trivial group names derived from unfamiliar words, CNIC recommended the following group names (*22f*), which also have not been widely accepted.

- Triels: B, Al, Ga, In, Tl (trielides in binary compounds)
- Tetrels: C, Si, Ge, Sn, Pb (tetrelides in binary compounds)
- Pentels: N, P, As, Sb, Bi (pentelides in binary compounds)

### 3.2.7 Elementlike Associations

The positively charged particles known as the positron ($e^+$) and muon ($\mu^+$) form short-lived associations with an electron that behave in some respects, including compound formation, like atoms of elements. These pseudoelement species [($e^+$)($e^-$) and ($\mu^+$)($e^-$)] are almost exclusively called positronium and muonium, respectively.

## 3.3 Polyhomoatomic Molecules

### 3.3.1 Unspecified Structure

An elementary substance that has an indefinite molecular structure or is a mixture of molecular species is given only the name of the element involved (examples: metallic copper is called copper and common sulfur is called sulfur).

### 3.3.2 Finite Structure

Neutral molecules consisting of two or more atoms of the same kind are named by placing in front of the name of the element the type-1 numerical prefix (*see* Table A.II; *see* Section 2.2.3 for discussion of different types of prefixes) corresponding to the number of atoms. The prefixes *cyclo-* and *catena-* may be used to designate rings and chains of atoms, respectively. Some examples are shown in Table 3.I.

**Table 3.I. Neutral Molecules**

| *Formula* | *Systematic Name* | *Trivial Name* |
|---|---|---|
| $O_2$ | Dioxygen | Oxygen |
| $O_3$ | *catena*-Trioxygen | Ozone |
| $S_8$ | *cyclo*-Octasulfur | α-, β-, or γ-Sulfur |
| $P_4$ | Tetraphosphorus | White phosphorus |

It is customary to omit di- in front of the element names for the common diatomic gaseous elements; however, the systematic names (such as dinitrogen and dioxygen) are used in recent British literature and are also used as ligand names (*see* Section 5.3.4.1). Because the numerical prefix di- is not generally used in the names for the gaseous forms of these elements, it is usual to refer to them in their atomic form as monoatomic hydrogen, monoatomic oxygen, etc.

The two kinds of molecular hydrogen, in which the nuclear spins of the two nuclei are parallel and opposed, are known as ortho and para hydrogen, respectively.

### 3.3.3 Allotropic Modifications

When the structure of an allotropic modification of an element is known, its name should indicate the molecular formula and/or crystal structure. *See* Section 3.3.2 for examples of names for molecular modifications. Crystalline allotropic modifications of an element may be named by adding in parentheses directly after the name of the element the Pearson symbol that gives the Bravais lattice (crystal class and type of unit cell) (*47*; *see* Table A.IV) in italics and the number of atoms in the unit cell for the structure of the allotrope. If the Pearson symbol fails to differentiate between two crystalline allotropes of an element, the symbol for the space group, separated by a comma and a space from the Pearson symbol, is added inside the parentheses, as shown in Table 3.II.

**Table 3.II. Crystalline Allotropes**

| *Formula* | *Systematic Name* |
|---|---|
| $C_n$ | Carbon(*cF*8) |
| | Carbon(*hP*4) |
| $Fe_n$ | Iron(*cI*2) |
| | Iron(*cF*4) |
| $P_n$ | Phosphorus(*oC*8) |
| $Se_n$ | Selenium(*mP*32, $P2_1/n$) |
| | Selenium(*mP*32, $P2_1/a$) |

Trivial names, generally names derived from the name of an element and a descriptor (color, Greek letter) or mineral names, continue to be accepted by CNIC as allowed alternatives to systematic names. Their extensive proliferation is discouraged, but such names are used for amorphous modifications of an element, for commonly recognized allotropes of indefinite structure, and for commonly occurring forms that are mixtures of structures or have ill-defined, disordered structures. Poly- is often used to signify "many" without specifying how many. Some examples of trivial names are shown in Table 3.III.

**Table 3.III. Trivial Names**

| *Formula* | *Trivial Name* |
|---|---|
| $C_n$ | Diamond [carbon(*cF*8)] |
| | Common graphite [carbon(*hP*4)] |
| | Vitreous carbon (amorphous carbon) |
| $Fe_n$ | α-Iron [iron(*cI*2)] |
| | γ-Iron [iron(*cF*4)] |
| $P_n$ | Black phorphorus [phosphorus(*oC*8)] |
| | Red phosphorus (amorphous phosphorus) |
| $Se_n$ | α-Selenium [selenium(*mP*32, $P2_1/n$)] |
| | β-Selenium [selenium(*mP*32, $P2_1/a$)] |
| $As_n$ | Amorphous arsenic |
| $S_8$ | γ-Sulfur (amorphous sulfur) |
| $S_n$ | μ-Sulfur (polysulfur) |

## 3.4 Homoatomic Ions

### 3.4.1 Monoatomic Cations

A positively charged monoatomic ion may be named by attaching in parentheses either the Stock number (*see* Section 2.3.2) or the Ewens–Bassett number (*see* Section 2.3.3) directly to the name of the element from which it is derived; that is, there is no space between the name and the parenthetical expression. The Ewens–Bassett number, which is the formal charge on the cation (or anion) expressed in terms of arabic numerals, is more generally applicable and is preferred by IUPAC. Often the word "cation" or "ion" is added as a separate word. Some examples are given in Table 3.IV.

**Table 3.IV. Monoatomic Cations**

| *Formula* | *Ewens–Bassett Name* | *Stock Name* |
|---|---|---|
| $Cr^{2+}$ | Chromium(2+) ion | Chromium(II) ion |
| $Cr^{3+}$ | Chromium(3+) ion | Chromium(III) ion |
| $Ag^{+}$ | Silver(1+) ion | Silver(I) ion |

The use of the suffixes -ous and -ic to indicate a lower and a higher charge on different cations derived from the same element, a use that was once almost universal, has all but disappeared. The Stock number or the Ewens–Bassett number may be omitted from the name when the element commonly forms only one cation (examples: sodium ion for $Na^{+}$ and calcium ion for $Ca^{2+}$).

### 3.4.2 Monoatomic Anions

A negatively charged monoatomic ion is named by adding the suffix "-ide" to a stem of the name of the element. Designation of the charge on monoatomic anions is usually omitted because elements seldom form more than one kind of monoatomic anion. Often the descriptive term "anion" or "ion" is added as a separate word even though it may be redundant. Some examples are shown in Table 3.V.

**Table 3.V. Monoatomic Anions**

| *Formula* | *Name* |
|---|---|
| $F^{-}$ | Fluoride ion |
| $Se^{2-}$ | Selenide ion |
| $P^{3-}$ | Phosphide anion |
| $Sn^{4-}$ | Stannide(4−) ion<br>Stannide(−IV) ion |

### 3.4.3 Polyhomoatomic Ions

An ion containing two or more atoms of the same kind is named by adding a type-1 numerical prefix (*see* Table A.II) to the cationic or anionic name for the element to indicate the number of atoms. The total amount of charge on the ion is specified by adding the Ewens–Bassett number parenthetically to the name of the ion. Although Stock numbers are also used, such usage is not preferred because there are a number of species that produce a fractional Stock notation (example: $O_2^{+}$). The separate words "ion", "cation", and "anion" are often added. Some polyhomoatomic ions have trivial names of long standing, which are included in the examples in Table 3.VI.

**Table 3.VI. Polyhomoatomic Ions**

| *Formula* | *Systematic Name* | *Trivial Name* |
|---|---|---|
| $Hg_2^{2+}$ | Dimercury(2+) ion[a] | Mercurous ion |
| $O_2^{+}$ | Dioxygen(1+) cation | Dioxygenyl(1+) |
| $Bi_5^{4+}$ | Pentabismuth(4+) | |
| $Ag_3^{+}$ | *cyclo*-Trisilver(1+) | |
| $I_4^{2+}$ | Tetraiodine(2+) | |
| $O_2^{2-}$ | Dioxide(2−) ion | Peroxide |
| $O_2^{-}$ | Dioxide(1−) ion | Hyperoxide<br>Superoxide |
| $C_2^{2-}$ | Dicarbide(2−) | Acetylide[b] |
| $N_3^{-}$ | Trinitride(1−) anion | Azide |
| $S_n^{2-}$ | Polysulfide(2−) | |
| $I_3^{-}$ | Triiodide(1−) ion | |

[a]The Stock name dimercury(I) cation is also possible.
[b]Acetylide has been used for both $C_2^{2-}$ and $HC_2^{-}$.

# Chapter 4
# Heteroatomic Species: General Principles

## 4.1 Introduction

Because the elements do not have systematic names, no procedure for inorganic nomenclature can be purely systematic beyond the indication of the elements present in a species and their structural arrangement. In general, heteroatomic species may be named in more than one way, and there is not necessarily a single "correct" name for each compound. To complicate matters further, the different systems of nomenclature are more or less continually changing as improvements are suggested and put into practice. Alternative and/or changing names are only a nuisance in the day-to-day practice of inorganic chemistry, but they pose a more serious problem in retrospective searching of the literature for more than a few recent years. This situation can be a particular problem as knowledge of a species increases and more informative names become prevalent. The use of trivial or semisystematic names in some of the examples in this book is not to be construed as an endorsement of such names, but just as a recognition that their use may sometimes be convenient.

Although the predominant systems for naming inorganic compounds are basically additive and the current tendency is to use additive nomenclature insofar as possible, other types of nomenclature are still used for some types of inorganic compounds (*see* Section 2.2.2). None of the systems in use is completely systematic, but all are systematic insofar as they provide formal procedures for assembling morphemes into names. The discussion of the nomenclature of heteroatomic species that follows is divided into chapters giving details for various procedures that have found some current or past use. Additive nomenclature is discussed as a general approach to inorganic nomenclature in Chapter 5, and then more specific procedures are considered in succeeding chapters. Several of the latter procedures are essentially embellishments to additive nomenclature.

A name can reproduce only the information that is known about the species to be named. Such information is generally available in a more or less detailed formula (*see* Section 2.4). It is important not to overinterpret a formula and draw unwarranted conclusions from it. A common mistake is to assign a coordination number on the basis of an empirical formula without taking into consideration the possibility of condensed structures. The discussion of individual systems in the chapters that follow proceeds from the use of systems that produce less informative names to systems that produce names giving more information about the compound being named.

## 4.2 Types of Heteroatomic Species

Groups of heteroatomic atoms may be neutral or ionic and may consist of few or many atoms. Neutral inorganic species are usually covalent molecules, but they are sometimes truly radicals in the physical organic chemistry sense, that is, neutral molecular structures containing one or more atoms carrying unpaired electrons (*48, 49*). The term "radical" has been used for a group of atoms the name of which is used as a prefix in substitutive nomenclature (*34*). Radicals in the latter sense usually do not have independent existence, and use of the term "radical" in this sense is now discouraged (*48, 49*). It would seem appropriate to make the same restriction in inorganic nomenclature. Furthermore, many inorganic species containing transition metals with unpaired electrons in inner electron shells meet the criterion for a radical species, but are not really chemically analogous to organic radicals.

The term "radical" will be restricted in this work to species with an unpaired electron in their outer electron shell, and the term "group" will be used for groups of atoms that are treated as a unit for nomenclature purposes. An unpaired outer electron may be indicated by placing a centered or superscript dot beside the appropriate atomic symbol in a structural formula or after a molecular formula.

In addition to the cations and anions that are found in ionic inorganic compounds, it is possible to have charged radicals under some conditions. The terms "radical cation" and "radical anion" are used for such species, and a superscript plus or minus sign is placed after the dot in the molecular formula according to inorganic (*30*) and organic (*34b*) nomenclature practices. Mass spectrometrists, however, place the superscript plus or minus sign in front of the superscript dot (*50*). The groups of atoms that are treated as units for inorganic nomenclature purposes are for the most part formally either cations or anions.

## 4.3 Heteroatomic Groups

### 4.3.1 General

In naming many compounds, it is customary to treat groups of atoms as single units for nomenclature purposes. The justification for doing this lies not in convenience alone but also in the observation that the same groups are present in large numbers of compounds and persist unchanged during reactions of these compounds.

This information is incorporated into inorganic nomenclature by giving such groups either trivial or systematic names and then naming their compounds as pseudobinary compounds. The subsections of Section 4.3 will be limited to the naming of relatively simple groups, but the principles therein are readily extended to complicated situations once a pattern has been established for the type of species under consideration.

Although the modern tendency is to use systematic names based on a central atom, many trivial names, for which there may be some systematic relationships, are still normally used. Binary groups for which a semisystematic nomenclature has evolved are of two types, those containing oxygen and another element (binary oxo groups) and those containing hydrogen and another element (binary hydro groups). The nomenclature of binary oxo groups will be discussed in this chapter in Section 4.3.3, whereas that of binary hydro groups will be included in the discussion of molecular hydrides as Section 9.3. The trivial name of cyanide for the anion $CN^-$ is also in common use and serves as the basis for naming $(CN)_2$, $CNO^-$, $CNS^-$, etc., which are treated as if $CN^-$ were a halide ion.

### 4.3.2 Additive Names

The additive name for a heteroatomic group of atoms is based upon the concept of a central atom to which other atoms are attached. If it is binary (that is, it contains atoms of only two elements), the procedure for giving it an additive name is straightforward. The atom of the more electropositive element according to Figure A.1 or the atom that appears only once in the group is considered to be the central atom, leaving the other atom(s) as the side atom(s). The name of the group is then formed by citing the appropriate morpheme for the side atom(s) with a multiplying prefix (*see* Table A.II), if necessary, followed by the name for the central atom. The morpheme used for the side atom(s) is a stem of the element name (*see* Table A.I) with the suffix "-o".

Examples:

| | |
|---|---|
| $SiH_3$ | trihydridosilicon |
| $CrO_2$ | dioxochromium |
| $CF_3$ | trifluorocarbon |
| $GeCl_3$ | trichlorogermanium |

An analogous procedure is used if the side atoms are not all atoms of the same element, with the morphemes for the side atoms cited in alphabetical order, ignoring any multiplying prefixes.

Examples:

| | |
|---|---|
| $PCl_2O_2$ | dichlorodioxophosphorus |
| $SClO_3$ | chlorotrioxosulfur |
| $SiBr_2Cl$ | dibromochlorosilicon |
| $SnClH_2$ | chlorodihydridotin |

Most heteroatomic groups do not occur as neutral groups, so the name derived in the foregoing fashion has to be modified to indicate that the group is present as a cation or an anion. The methods for doing this are discussed in Section 4.5.2.

Examples:

| | |
|---|---|
| $UO_2^{2+}$ | dioxouranium(2+) cation |
| $VO_4^{3-}$ | tetraoxovanadate(3−) anion |

If, instead of single atoms, there are one or more groups of atoms bonded to a central atom, the names for these side groups are formed in the usual way for groups of atoms, adding the suffix "-o" or "-ato" where appropriate. Organic side groups are named as hydrocarbon groups (or substituted derivatives) (*22b*) or by modification of the organic name with one of these suffixes.

Examples:

| | |
|---|---|
| $SiCl(OSiCl_3)_2$ | chlorobis(trichlorooxosilicato)silicon |
| $Sn(CH_3)_3$ | trimethyltin |
| $Pb(OOCCH_3)Br_2$ | (acetato)dibromolead |

### 4.3.3 Binary Oxo Groups

Names for heteroatomic groups consisting of a characteristic element E and one or two oxygen atoms may be formed by adding "-osyl" or "-yl", respectively, as a suffix to a stem name for E. For similar groups with three oxygen atoms, the prefix "per-" may be added to the "-yl" names for the corresponding $EO_2$ groups. These names refer to uncharged groups of atoms and require modification in order to designate cationic or anionic groups (*see* Section 4.3.2).

Examples:

| EO | | $EO_2$ | | $EO_3$ | |
|---|---|---|---|---|---|
| ClO | chlorosyl | $ClO_2$ | chloryl | $ClO_3$ | perchloryl |
| IO | iodosyl | $IO_2$ | iodyl | $IO_3$ | periodyl |
| NO | nitrosyl | $NO_2$ | nitryl | | |
| | | $CrO_2$ | chromyl | | |
| | | $UO_2$ | uranyl | | |
| | | | | $MnO_3$ | permanganyl |
| | | | | $ReO_3$ | perrhenyl |
| | | | | $TcO_3$ | pertechnetyl |

The EO groups in the following list have been known for a long time by the names that would be given to the corresponding $EO_2$ groups by the preceding method. Although each EO group could be renamed according to that method as an -osyl group, the -yl name cannot be used for the corresponding $EO_2$ group because of the consequent confusion. Thus, they are exceptions to the method.

Exceptions:

| | | | |
|---|---|---|---|
| BiO | bismuthyl | TiO | titanyl |
| SbO | antimonyl | PO | phosphoryl |
| CO | carbonyl | VO | vanadyl |

A number of EO and $EO_2$ groups that could be named by the semisystematic method are commonly known by other names that presumably will not disappear quickly. If that method is applied to the groups in question, however, the stem name for the characteristic element must be chosen carefully to avoid formation of a name used for another group. For example, $SeO_2$ must be named selyl because selenyl is a name for HSe (*see* Section 9.3). Trivial names for EO and $EO_2$ groups are

| | | | |
|---|---|---|---|
| HO | hydroxyl | SeO | seleninyl |
| $HO_2$ | perhydroxyl | $SeO_2$ | selenonyl |
| SO | sulfinyl[a] | $PO_2$ | phospho |
| $SO_2$ | sulfonyl[b] | $AsO_2$ | arso |
| | | $SbO_2$ | stibo |

[a]Thionyl was commonly used. The semisystematic method gives sulfurosyl.

[b]Sulfuryl, a name formerly in common use, is the name formed according to the semisystematic method.

Chalcogen analogs of binary oxo groups may be named by citing the prefixes "thio-", "seleno-", and "telluro-", in alphabetical order when two or more are present, in front of the name of the corresponding binary oxo group. Type-1 numerical prefixes (*see* Table A.II), with the exception of mono, are used as appropriate.

Examples:

| | | | |
|---|---|---|---|
| NS | thionitrosyl | CSe | selenocarbonyl |
| $PS_2$ | dithiophospho | ISSeO | selenothioperiodyl |

## 4.4 Stoichiometric Names

### 4.4.1 Numerical Prefixes

When only the empirical formula for a compound is known, a name denoting the stoichiometric proportions is formed by joining type-1 numerical prefixes (*see* Table A.II) without hyphens to names or modified names for the elements comprising the compound. The end vowels of the prefixes should not be elided. The prefix mono may be omitted except when confusion is possible. Arabic numerals with hyphens may be used instead of prefixes for numbers greater than 10 in order to facilitate understanding, but numbers are much less desirable than prefixes. Alphabetical listings, for example, are much more difficult to understand if numbers are used.

The order for citing the names of the elements in a binary compound is determined by their relative electronegativities, as given in Figure A.1. The name used for the more electropositive element is the unmodified name of that element, as given in Table A.I. It is cited first, followed by the name for the more electronegative element, which is formed by adding the suffix "-ide" to a stem of that element's name (*see* Table A.I). When more than two elements are present, the most electropositive and the most electronegative are named as if they are constituents of a binary compound. The remaining elements in the compound are cited by their unmodified names if they are regarded as electropositive and by the name derived by adding the suffix "-ide" to the appropriate stem of their name if they are regarded as electronegative. All unmodified names are cited in one alphabetical sequence, ignoring any numerical prefixes, and all names ending with the suffix "-ide" are cited in a second alphabetical sequence, again ignoring any numerical prefixes. Names formed by the methods of this paragraph are used primarily for binary compounds because it is generally possible to use more informative names for compounds containing more than two elements.

Examples:

| | |
|---|---|
| $NO_2$ | nitrogen dioxide |
| $FeCl_3$ | iron trichloride |
| $N_2O_5$ | dinitrogen pentaoxide |
| $UCl_3F$ | uranium trichloride fluoride |
| $Ca_3P_2O_8$ | tricalcium diphosphorus octaoxide |
| $K_2PtCl_6$ | platinum dipotassium hexachloride |
| LiH | lithium hydride |
| $Al_3Fe$ | iron trialuminide |
| $BrCl_3$ | bromine trichloride |

### 4.4.2 Stock Notation

Stoichiometric composition may be indicated indirectly by using the Stock notation for oxidation number (*see* Section 2.3.2), which may be applied to both electropositive and electronegative elements.

Examples:

| | |
|---|---|
| $MnO_2$ | manganese(IV) oxide |
| $P_2O_5$ | phosphorus(V) oxide |
| $FeCl_3$ | iron(III) chloride |
| $FeCl_2$ | iron(II) chloride |

| | |
|---|---|
| $Na_2CoI_4$ | cobalt(II) disodium iodide |
| HO | hydrogen oxide(–I) |
| $K_2PtCl_6$ | platinum(IV) dipotassium chloride |

Prior to the introduction of the Stock notation, when an element exhibited more than one oxidation number, suffixes, prefixes, or both were used to differentiate among oxidation numbers. If an element exhibited two oxidation numbers, the suffix "-ic" was used to designate the higher oxidation number and the suffix "-ous" the lower; for example, ferric is used for Fe(III), ferrous for Fe(II), phosphoric for P(V), and phosphorous for P(III). The prefixes "per-" and "hypo-" were combined with these suffixes to indicate other oxidation numbers when there were more than two. For the most part this system is no longer in use, but *see* Section 8.2.3 for its use with certain oxo acids (and their derivatives). It has also been retained in indicating that the oxidation number of oxygen is not the usual 2–; the name for $O_2^{2-}$ is peroxide anion and for $O_2^-$ is hyperoxide anion.

### 4.4.3 Ewens–Bassett Notation

The Ewens–Bassett notation (*see* Section 2.3.3) may be used in the same way as the Stock notation (*see* Section 4.4.2) to indicate the stoichiometric composition indirectly. This method is particularly useful for binary compounds when it is desired to emphasize the presence of discrete ions.

| Examples: | | |
|---|---|---|
| | AgF | silver(1+) fluoride |
| | $AgF_2$ | silver(2+) fluoride |
| | $Na_2S$ | sodium sulfide(2–) |
| | $Fe_3O_4$ | iron(2+) diiron(3+) oxide |
| | $TiCl_3$ | titanium(3+) chloride |
| | $K_2PtCl_6$ | potassium hexachloroplatinate(2–) |

## 4.5 Names Indicating Atomic Groupings

### 4.5.1 Discrete Molecules

When the molecular formula of a compound is known, this information can be indicated by citing the actual number of atoms or groups of atoms of each kind in a molecule of the compound, rather than by simply giving the stoichiometry as described in Section 4.4. Type-1 or type-2 numerical prefixes (*see* Table A.II) are used as appropriate. If the molecular formula and the stoichiometric formula are the same, the molecular name will, of course, be the same as the stoichiometric name.

| Examples: | | |
|---|---|---|
| | $N_2O_4$ | dinitrogen tetraoxide |
| | $S_2Cl_2$ | disulfur dichloride |
| | $P_4S_{10}$ | tetraphosphorus decasulfide |
| | HCN | hydrogen cyanide |
| | $(ClO_3)F$ | perchloryl fluoride |

### 4.5.2 Ions

The naming of homoatomic cations and anions is discussed in Section 3.4. Heteroatomic ions that have the same elemental composition as the heteroatomic groups discussed in

Section 4.3 may be named by adding the appropriate class term ("anion" or "cation") as a separate word after the name of the group. The amount of charge may be indicated by adding a type-1 numerical prefix (*see* Table A.II) to the class term or by adding either the Stock number of the characteristic element in parentheses or the Ewens–Bassett number in parentheses to the name of the group. In the last case, the term "ion" is usually used instead of "anion" or "cation".

Examples:

| | |
|---|---|
| $[NO]^+$ | nitrosyl cation |
| $[NO_2]^-$ | nitryl anion |
| $[CO]^{2+}$ | carbonyl dication<br>carbonyl(2+) ion |
| $[PO]^{3+}$ | phosphoryl trication<br>phosphoryl(V) cation<br>phosphoryl(3+) ion |
| $[VO]^{2+}$ | vanadyl dication<br>vanadyl(IV) cation<br>vanadyl(2+) ion |
| $[SeO_3S]^{2-}$ | thioselenate dianion<br>thioselenate(2−) ion |
| $[UOS]^{2+}$ | thiouranyl dication<br>thiouranyl(VI) cation<br>thiouranyl(2+) ion |

The traditional nomenclature of the many binary oxo anions formally derived from oxo acids by the loss of one or more hydrogen(1+) ions is discussed in Chapter 8. For example, the traditional name for $SO_4^{2-}$ is sulfate. However, such anions also may be named by applying the principles in the preceding paragraph to additive names formed according to Section 4.3.2.

Examples:

| | |
|---|---|
| $[SO_4]^{2-}$ | tetraoxosulfate dianion<br>tetraoxosulfate(VI) anion<br>tetraoxosulfate(2−) ion |
| $[NbO_3]^-$ | trioxoniobate anion<br>trioxoniobate(V) anion<br>trioxoniobate(1−) ion |
| $[SeO_3S]^{2-}$ | trioxothioselenate dianion<br>trioxothioselenate(VI) anion |
| $[SbS_4]^{3-}$ | tetrathioantimonate trianion<br>tetrathioantimonate(V) anion<br>tetrathioantimonate(2−) ion |

Binary ions that can be derived formally by the addition or removal of hydrogen ions from a molecular hydride may be named by the method given in Section 9.4, as well as by the method of this section. *See* Section 9.4 for examples of names for binary hydro ions derived by both methods.

### 4.5.3 Ionic Compounds

Compounds containing polyatomic ions may be named by citing cations before anions and putting each class in alphabetical order when more than one cation, anion, or both are present. Such names give information about how some atoms are associated in a compound, as well as giving the stoichiometry. The naming of ions is discussed in Sections 3.4, 4.5.2, 8.4, and 9.4. Table A.V is a compilation of trivial and semisystematic names for parent polyatomic ions that are not covered by systematic nomenclature or the conventions for binary oxo and hydro groups.

Examples:

| | |
|---|---|
| $Cu(SO_4)$ | copper sulfate<br>copper(II) sulfate<br>copper(2+) sulfate |
| $Ni(CN)_2$ | nickel dicyanide<br>nickel(II) cyanide<br>nickel(2+) cyanide |
| $(NH_4)_4[Fe(CN)_6]$ | tetraammonium ferrocyanide<br>ammonium ferrocyanide(4−) |
| $Na_3(PO_4)$ | trisodium phosphate<br>sodium phosphate(3−) |
| $K(PO_3)$ | potassium metaphosphate(1−) |
| $(Hg_2)Cl_2$ | dimercury dichloride<br>dimercury(2+) chloride |
| $(UO_2)(NO_3)_2$ | uranyl dinitrate<br>uranyl(2+) nitrate |

### 4.5.4 Radicals and Radical Ions

Inorganic species that meet the criterion for a radical, known earlier as a "free radical" or a radical ion (*32a, 34*), that is, that have unpaired electrons in the outermost electron shells of one or more of the atoms making up the species (*see* Section 4.2), can always be named as discrete molecules (*see* Section 4.5.1) or as ions (*see* Section 4.5.2). When it is desired to emphasize that the species is a radical, the words radical or radical ion can be added to the name for the species.

Examples:

| | |
|---|---|
| ClO | chlorine(II) oxide radical<br>chlorosyl radical |
| HO | hydroxyl radical |
| $SF_4^+$ | tetrafluorosulfur radical ion(1+) |
| $NO_2^{2-}$ | nitryl radical ion(2−) |

## 4.6 Names Indicating Structure

### 4.6.1 Discussion

Inorganic compounds may be simple molecules, aggregates of discrete ions, or large molecules consisting of infinite chains, sheets, or three-dimensional networks. Systems of nomenclature have been designed to make it possible to indicate the stereochemical structure for many of these possibilities. Often the structure for a simple molecule or discrete ion is assumed by analogy to known structures, and it is not considered necessary to include extensive stereochemical information in the name. In this section methods for differentiating isomers will be mentioned, but the detailed treatment of stereochemical nomenclature will be presented separately in Chapter 16.

### 4.6.2 Association of Atoms

The manner in which atoms are bonded one to another in a molecule is the most elementary structural information. It does not provide geometrical information, although in many cases the geometry of a molecule is considered to be evident on the basis of theoretical considerations once the number of bonds formed by each element is known. Many of the names formed according to the principles of Section 4.5 give bonding information. A systematic procedure for constructing names that indicate how the atoms in a molecule are joined has been developed from the method for naming coordination compounds suggested by Werner (*11*). This system will be described in detail in Chapter 5. The following examples are included only as simple illustrations of the method. Names based on Stock numbers can also be used.

Examples:

| | |
|---|---|
| $K_2SO_4$ | potassium tetraoxosulfate(2−)<br>potassium tetraoxosulfate(VI) |
| $Ca_2[Fe(CN)_6]$ | calcium hexacyanoferrate(4−) |
| $[Cu(NH_3)_4][Ni(NCS)_4]$ | tetraamminecopper(2+) tetrakis(thiocyanato)nickelate(2−) |

### 4.6.3 Stereochemistry

If the names for isomers are otherwise identical, that is, if the species in question are stereoisomers, prefixes may be used to designate specific isomers in simple cases (*see* Section 5.3.6.3). Locants have been used in more complex cases (*see* Section 16.2), but the most recent procedure has been to designate the detailed spatial relationship of the individual atoms by the use of more or less complex descriptors entailing the use of italicized letters and arabic numerals (*see* Section 16.3).

# Chapter 5
# Additive Nomenclature

## 5.1 Introduction

The procedure for naming coordination entities, which is a refinement and extension of the system proposed by Werner (*11*), was mentioned earlier as an example of the additive method of nomenclature. A coordination entity is defined as a group of atoms that contains one or more coordinate covalent bonds and retains its identity during metathetical chemical reactions. It may be a neutral molecule, a cation, or an anion that is considered to consist of a central atom (also called a nuclear atom or center of coordination), to which side groups known as ligands are bonded. The ligands, which may be single atoms or groups of atoms, may be viewed conceptually as neutral or ionic entities that are bonded to an appropriately charged central atom. Although the method for naming coordination entities is a systematic additive procedure, the names produced are not always purely systematic because the morphemes incorporated into the final name may not be systematic.

This chapter will begin with a discussion of naming mononuclear coordination entities, that is, coordination entities with only one central atom. Chapters 6, 7, and 16, which are devoted to naming polynuclear coordination entities, naming inorganic polymers, and indicating the stereochemistry of inorganic compounds, respectively, extend the procedure for naming mononuclear coordination entities in various ways.

The method of this chapter is not dependent on the presence of coordinate covalent bonds and hence is applicable to any group of atoms, neutral or charged, in which there is a central atom surrounded by two or more atoms or groups of atoms. This chapter will end with a discussion of the general use of additive nomenclature for inorganic species.

## 5.2 Choice of Central Atom

The central atom (ion) for a conventional mononuclear coordination entity based upon a metallic element is the atom (ion) of that metal. However, specific rules are needed for the choice of a central atom when additive nomenclature is applied to compounds containing two or more nonmetallic atoms in higher oxidation states. The following order of priorities, listed in decreasing order of preference, is suggested. The symbol $>$ indicates "has preference over what follows".

1. Metals $>$ nonmetals (except carbon) $>$ carbon
2. Most centrally located atom [example: S of $SO_2$ in preference to S of either $SF_5$ in $F_5SOS(O)_2OSF_5$]

3. Element encountered last when going from electronegative to electropositive elements according to Figure A.1 [example: P in preference to S in $FS(O)_2OP(O)F_2$]
4. Element with higher oxidation number [example: S(VI) of $SF_5$ in preference to S(IV) of FS(O) in $FS(O)OSF_5$]
5. Element with greater number of side groups [example: S of $SF_5$ in preference to S of $S(O)_2F$ in $F_5SOS(O)_2F$]
6. Atom with greater variety of bonding atoms (example: S of $SClF_4$ with bonds to Cl, F, and O in preference to S in $SF_5$ with bonds to only F and O in $F_5SOSClF_4$)
7. Atom to which is attached the ligand or side group with the name first in alphabetical order [example: S of $S(O)_2Cl$ (chloro side group) in preference to S of $FS(O)_2$ (fluoro side group) in $FS(O)_2OS(O)_2Cl$]

## 5.3 Naming a Coordination Entity

### 5.3.1 General Procedure

A coordination entity is named by citing the names of the ligands in alphabetical order, followed by an appropriate name for the central atom, with an indication of the charge on the coordination entity or the oxidation state of the central atom. In addition to these essentials, descriptor prefixes or symbols may be used to indicate details of structure. If the coordination entity is an ion, any compound containing that ion is named as an ionic compound (*see* Section 4.5.3). In formulas the coordination entity is usually enclosed in brackets (*see* Section 2.2.7). These principles are intrinsic to the remainder of this chapter, so examples will not be given at this point.

The order of citation of the ligands has not always been alphabetical, and various orders have been used in the past. Werner's practice of citing anionic ligands before neutral ligands is reflected in the 1940 Inorganic Nomenclature Rules (*19*), which were inadequate for the order within these groups as the variety of ligands increased. The 1957 IUPAC Inorganic Rules (*20, 21*) gave an order of priorities among anionic ligands that was based essentially on first the number of atoms constituting inorganic anions and then the atomic number of the central atom in the ion, followed by organic anions in alphabetical order. The order among neutral ligands was given as water, ammonia, other inorganic ligands in a sequence based roughly on the position of their coordinating atoms in the periodic table, and finally organic ligands in alphabetical order. The complexity of the rules combined with the ever-increasing variety of ligands finally led to the decision in 1970 to use a single series of alphabetically arranged ligand names without regard to ligand type (*22i*).

### 5.3.2 Central Atom

The name for the central atom in a neutral or positively charged coordination entity is the name of the element. The name for the central atom in a negatively charged coordination entity is formed by adding the suffix "ate" to an element stem name for the central atom (*see* Table A.I).

### 5.3.3 Charge on Coordination Entity

The charge on ionic coordination entities can be indicated directly with the Ewens–Bassett number (*see* Section 2.3.3) or indirectly by indicating the oxidation number of the central atom with the Stock number (*see* Section 2.3.2). The use in the Stock system of 0 for an oxidation number of zero (because there is no Roman numeral for zero) precludes the use of 0 as a Ewens–Bassett number for neutral coordination entities. Normally this causes no confusion.

After Stock proposed the use of parenthetical numbers to indicate the oxidation number of the central element (*see* Section 2.3.2), the adoption of Werner's proposals for naming coordination compounds was rapid. Until then there had been some reluctance to use the Werner system because of its awkward way of indicating oxidation numbers. Even the Stock proposal was not entirely satisfactory. First, the charge on the coordination entity is not immediately evident from the Stock number but must be calculated. Second, there are many instances in which the oxidation number of the central atom is not at all evident, and, in fact, it may have no real meaning. Although both systems are widely used, the Ewens–Bassett number is generally more satisfactory and will be used in most examples in this book with the understanding that in most cases Stock numbers would serve as well.

### 5.3.4 Ligand Names

***5.3.4.1 Neutral Ligands.*** The name for a neutral molecule is also used as the name for that molecule when it functions as a ligand, except for those simple molecules listed in this section. The numerical prefix di- is included in the names for diatomic elements acting as ligands, even though that prefix is usually omitted from the names for the elements themselves (*see* Section 3.3.2) (example: dinitrogen). If a ligand is customarily named by binary nomenclature, the separation of words is maintained in the name of the coordination entity. The ligand name is set off with enclosing marks, except for the names of the simple molecules that follow.

**Ligand Names Not Set Off with Enclosing Marks**

| | | | |
|---|---|---|---|
| $C_2H_4$ | ethylene | $H_2O$ | aqua |
| CO | carbonyl | $NH_3$ | ammine |
| CS | thiocarbonyl | NO | nitrosyl |
| | | NS | thionitrosyl |

***5.3.4.2 Cationic Ligands.*** The name for a cationic entity functioning as a ligand is the same as the name for the cation (examples: oxonium, $\eta^7$-cycloheptatrienylium).

***5.3.4.3 Anionic Ligands.*** The name for an anionic entity functioning as a ligand is generally formed by modifying the name of the anion so that it ends with "-o". The common suffixes of anion names, "-ide", "-ate", and "-ite", are changed to "-ido", "-ato", and "-ito", respectively, in ligand names (examples: amido, acetato, methyl sulfito). There are some exceptions in which "-ide" is changed to "-o", namely, bromo ($Br^-$), chloro ($Cl^-$), cyano ($CN^-$), fluoro ($F^-$), hydrogen peroxo* ($HO_2^-$), hydroxo* ($OH^-$), iodo ($I^-$), mercapto or hydrosulfido ($HS^-$), oxo* ($O^{2-}$), peroxo* ($O_2^{2-}$), and thio* ($S^{2-}$). Anionic ligands derived from amines, amides, and imines are named as substituted amido or imido ligands {examples: methylamido for $CH_3NH^-$, acetylamido (usually contracted to acetamido) for $CH_3CONH^-$, and ethylimido for $CH_3CH_2N^{2-}$}.

The phosphorus and arsenic analogs may be named additively, based on the names phosphido and arsenido, or substitutively, based on the names phosphanido and arsanido

*In CAS index nomenclature the names hydro (instead of hydrido for $H^-$), hydroperoxy, hydroxy, oxy, peroxy, and thioxo are used.

{example: diphenylphosphido or diphenylphosphanido for $(C_6H_5)_2P^-$}. Anionic ligands derived from alcohols and their sulfur, selenium, and tellurium analogs are named by adding "-ato" to "-ol", "-thiol", etc. (examples: methanolato for $CH_3O^-$ and benzenethiolato for $C_6H_5S^-$). The use of phenylmercapto, etc., is not approved by nomenclature committees, although such usage may be encountered. An organic ligand that can be viewed as a hydrocarbon group bonded to a central atom is given the appropriate organic name without alteration (example: ethyl for $C_2H_5$) (*34*). All other anionic organic ligands are named by adding "-ato" to the organic name of the parent compound, eliding the final "-e" if necessary {example: 2,4-pentanedionato for $[H_3CC(O)CHC(O)CH_3]^-$}. (*See also* the naming of organometallic compounds, Chapter 12.)

Some side groups are given names different from ligand names in substitutive nomenclature (*see* Section 9.2). Table A.V gives both names for many side groups.

### 5.3.5 Indication of Ligating Atom(s)

*5.3.5.1* ***Italicized Element Symbols.*** The ligating atom in an ambidentate ligand or the particular ligating atoms in a flexidentate ligand may be indicated by adding a hyphen followed by the italicized atomic symbol(s) for the ligating atom(s) to the name of the ligand. Thus, $CN^-$ can bond either via the C (cyano-*C*) or the N (cyano-*N*), and $NO_2^-$ can bond either via the N (nitro-*N*) or an O (nitro-*O*). Application of this method to complicated ligands requires the use of superscript locants and primes and a set of priorities for ordering the ligand locants. For example, 1,2-ethanediyldinitrilotetraacetato-*O,O* could mean bonding with two O atoms of one COO group, one O from each of the $CH_2COO$ groups bonded to a single N, or one O from each of two $CH_2COO$ groups bonded to different N atoms. A special name has been given to each form of an ambidentate ligand in a few cases (examples: cyanato for cyanato-*O* and isocyanato for cyanato-*N*; nitrito for nitro-*O* and nitro for nitro-*N*), but such a system is difficult to generalize.

Conventions in the 1970 IUPAC Inorganic Nomenclature Rules (*22j*) for selecting the order in which to list the symbols for ligating atoms, which were designed for correlation with positions on coordination polyhedrons, are complex. They have been used only rarely. However, the principle of using italicized element symbols to identify ligating atoms in complicated ligands can be used by listing the element symbols of the ligating atoms alphabetically by symbol, differentiating among like atoms by serial primes or superscripts derived from the numbering of the ligand based on organic nomenclature practices (*34c*). It may be necessary to provide additional superscript characters when none are readily apparent from organic nomenclature principles.

Examples (ligating atoms are marked *):

| | |
|---|---|
| $H_2N^*CH(CH_3)CH_2N^*H_2$ | 1,2-propanediamine-*N,N'* |
| $[CF_3C(O^*)CHC(O^*)CH_3]^-$ | 1,1,1-trifluoro-2,4-pentanedionato-*O,O'* |
| $H_2N^*CH_2CH_2S^*CH_3$ | 2-(methylthio)ethylamine-*N,S* |
| $[O^*OCCH_2N^*H_2]^-$ | glycinato-*N,O* |
| $[OO^*\underset{1}{C}\underset{2}{C}H_2(O^*)\underset{3}{C}H(OH)\underset{4}{C}OO]^{3-}$ | tartrato(3–)-$O^1,O^2$ |
| $[OO\underset{1}{C}\underset{2}{C}H_2(O^*)\underset{3}{C}H(O^*)\underset{4}{C}OO]^{4-}$ | tartrato(4–)-$O^2,O^3$ |
| $H_2N^*CH_2CH_2S^*CH_2CH_2N(CH_3)_2$ | *N,N*-dimethyl-2,2'-thiobis(ethylamine)-*N',S* |
| (structure below) | 1,3-dithiolane-2-thione-$S^1,S^2$ |

```
S*    S*
  \\C2 / 1 \ 5CH2
    \3      4/
     S------C
            H2
```

| | |
|---|---|
| $[H_2NCH_2C(O)N^*H\underset{2}{C}H(\underset{3}{C}H_3)\underset{1}{C}OO^*]^-$ | *N*-glycylalaninato-$N^2,O^1$ |
| $[H_2N^*CH_2C(O^*)NHCH(CH_3)COO]^-$ | *N*-glycylalaninato-$N^N,O^N$ |

*5.3.5.2 The Kappa Convention.* As the complexity of the ligand name increases, the utility of the approach in Section 5.3.5.1 decreases because the same atomic element indexes are used as locants for organic substituents within the name of the ligand itself. The potential confusion is avoided in the kappa convention (*51*) by placing the Greek letter κ before the italicized element symbol for the ligating atom. The ligand locant is placed after that portion of the ligand name to which it directly applies, such as the part of the ligand name denoting the chain, ring, or group to which the ligating atom is bound. The position of the ligating atom in the ligand is indicated by the use of superscript numerals, letters, or primes on the element symbol. For multidentate ligands a superscript number may be given on κ to indicate the number of identically bound ligating atoms in the flexidentate ligand.

Examples (ligating atoms are marked *):

| | |
|---|---|
| $(NCS^*)^-$ | thiocyanato-$\kappa S$ |
| $(Ph)_3P^*$ | triphenylphosphine-$\kappa P$ |
| $H_2N^*CH_2CH_2N^*H_2$ | 1,2-ethanediamine-$\kappa^2N,N'$ (also ethylenediamine-$\kappa^2N,N'$) |
| $H_2N^*C_2H_4N^*HC_2H_4NHC_2H_4N^*H_2$ | *N,N'*-bis(2-amino-$\kappa N$-ethyl)-1,2-ethanediamine-$\kappa N$ (The nonsystematic name diethylenetriamine-$\kappa^3N,N',N'''$ is also used.) |
| $H_2N^*C_2H_4N^*HC_2H_4N^*HC_2H_4NH_2$ | *N*-(2-amino-$\kappa N$-ethyl)-*N'*-(2-aminoethyl)-1,2-ethanediamine-$\kappa^2N,N'$ (The nonsystematic name diethylenetriamine-$\kappa^3N,N',N''$ is also used.) |
| $[(OOCCH_2)_2N^*C_2H_4N^*(CH_2COO)_2]^{4-}$ | 1,2-ethanediyldinitrilo-$\kappa^2N,N'$-tetraacetato(4–) (also ethylenedinitrilo-$\kappa^2N,N'$-tetraacetato(4–), ethylenediamine-$\kappa^2N,N'$-tetraacetato(4–), or any of these names with tetrakisacetato in place of tetraacetato) |
| $[(O^*OCCH_2)(OOCCH_2)N^*C_2H_4N(CH_2COO)_2]^{4-}$ | 1,2-ethanediyldinitrilo-$\kappa N$-tetraacetato(4–)-$\kappa O$ (also all alternatives in analogy to the preceding example) |
| $[(O^*OCCH_2)(OOCCH_2)N^*C_2H_4N^*(CH_2COO^*)(CH_2COO)]^{4-}$ | 1,2-ethanediyldinitrilo-$\kappa^2N,N'$-tetraacetato(4–)-$\kappa^2O,O'$ (also all analogous alternatives as above) |
| $PhC(=N^*H)NHS(=N^*C_2H_4N^*H_2)(=NPh)Ph$ | *N*-[*N*-(2-amino-$\kappa N$-ethyl)-*N'*,*S*-diphenylsulfonodiimidoyl-$\kappa N$]benzamidine-$\kappa N'$ |
| $PhC(=NH)NHS(=N^*C_2H_4N^*H_2)(=N^*Ph)Ph$ | *N*-[*N*-(2-amino-$\kappa N$-ethyl)-*N'*,*S*-diphenylsulfonodiimidoyl-$\kappa^2N,N'$]benzamidine |
| $PhC(=NH)N^*HS(=N^*C_2H_4N^*H_2)(=NPh)Ph$ | *N*-[*N*-(2-amino-$\kappa N$-ethyl)-*N'*,*S*-diphenylsulfonodiimidoyl-$\kappa N$]benzamidine-$\kappa N$ |

*5.3.5.3 The Hapto Convention.* When two or more carbon atoms from a chain or ring are bonded to a central atom, this may be indicated by the symbol $\eta^n$ (*22k, 52*), in which *n* is the number of carbon atoms involved in the bonding, prefixed to the ligand name or that portion of the name most appropriate to indicate the connectivity. Locants may be used to indicate the specific ligating atoms involved. When locants are used, they and η are enclosed in parentheses, but the superscript to η is omitted because it is superfluous. The use of η will be considered further in the discussion of organometallic compounds in Chapter 12.

Examples:

| | |
|---|---|
| $H_2C{=}CH_2$ | $\eta^2$-ethene<br>$\eta^2$-ethylene |
| $[H_2C{=}CHCH_2]^-$ | $\eta^2$-allyl |
| $C_5H_5^-$ | $\eta^5$-cyclopentadienyl |

| | |
|---|---|
| $[H_2C{=}CHC_5H_4]^-$ | ($\eta^2$-ethenyl)cyclopentadienyl (for bonding via the $CH_2{=}CH$ group)<br>ethenyl-$\eta^5$-cyclopentadienyl (for bonding via the $C_5H_4$ ring) |
| 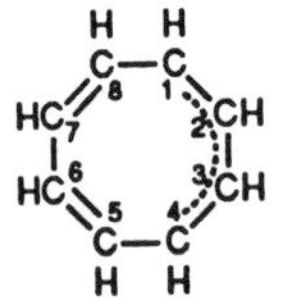 (cyclooctatetraene structure, atoms numbered 1–8) | (1,2,3,4-$\eta$)-1,3,5,7-cyclooctatetraene |

## 5.3.6 Spatial Relationships

*5.3.6.1 General Comments.* Associated with each central atom is at least one spatial distribution of the linkages from it to the side groups. Each distribution can be identified by the name of a geometrical figure, the apexes of which are occupied by the bonding atoms (and the central atom when it is not enclosed by the bonding atoms). The great majority of inorganic compounds normally encountered are of a relatively few configurational types, so there has been no pressing need to designate the configurations in the name. At first the greatest need was to be able to distinguish between tetrahedral and square (usually given the redundant name "square planar") configurations, especially when a central atom, such as nickel, could exhibit both. The increasing number of compounds exhibiting coordination numbers other than 4 or 6, along with the discovery of a variety of configurations for coordination numbers other than 4, has led to the development of methods for the succinct indication of these various configurations. The complete stereochemical description of inorganic compounds will not be considered in this chapter, but will be discussed fully in a later chapter on stereochemistry (*see* Chapter 16).

*5.3.6.2 Designation of Configuration.* For nearly half a century, almost all coordination compounds studied exhibited a coordination number of either 4 or 6, as deduced from stoichiometry. A coordination number of 6 seemed invariably to correspond to an octahedral configuration, and those who studied coordination compounds learned to recognize instinctively whether a four-coordinate compound was square or tetrahedral. This intuitive recognition of the configuration of a compound has continued, and for many purposes the configuration is still not indicated in the name. There is, however, no reason why symmetry site terms (*see* Section 16.3.2) should not be used by themselves in names when designation of the configuration is desired but a complete stereochemical description is not needed [example: *OC*-6-hexaamminecobalt(3+) chloride].

*5.3.6.3 Location of Side Groups.* Prefixes have long been used to designate relationships of ligands one to another. The prefixes in use are:

- *cis-* to designate two ligands occupying adjacent positions,
- *trans-* to indicate two ligands occupying positions at the ends of a major axis,
- *fac-* to indicate three ligands occupying the three positions on a trigonal face of a polyhedron, and
- *mer-* to indicate three ligands occupying adjacent positions on an edge of a polyhedron.

A more precise method for locating ligands is to assign a locant to each ligand, as described in Section 16.2.2.

*5.3.6.4 Designation of Chirality.* Before absolute configurations were established, it was customary to distinguish stereoisomers by the use of the descriptors *d* (*dextro*, +)

and *l* (*levo*, –), which indicated how the individual isomers affected polarized light. Now the chirality or asymmetrical arrangement of ligating atoms about a central atom may be indicated by a configuration index that is assigned according to conventions outlined in Chapter 16. This aspect of naming coordination compounds will be deferred to that discussion.

## 5.3.7 Examples

*5.3.7.1 Names Indicating Composition.* The least informative additive name that can be given to a coordination entity merely specifies the central atom, the ligands, and the charge on the entity. The use of Stock numbers instead of the Ewens–Bassett numbers shown in the examples is acceptable.

| | |
|---|---|
| $[Pt(NH_3)_6]^{4+}$ | hexaammineplatinum(4+) ion |
| $[Cr(CN)_6]^{3-}$ | hexacyanochromate(3–) ion |
| $[Be(AcCHAc)_2]$ | bis(2,4-pentanedionato)beryllium |
| $[Co(H_2NC_2H_4NH_2)_2(CO_3)]^+$ | (carbonato)bis(1,2-ethanediamine)cobalt(1+) ion |
| $[Pt\{H_2NCH_2CH(NH_2)CH_2NH_3\}Cl_2]^+$ | dichloro[(2,3-diaminopropyl)ammonium]platinum(1+) ion |
| $[Co(N_3)(NH_3)_5]SO_4$ | pentaammineazidocobalt(2+) sulfate |

*5.3.7.2 Names Indicating Bonding or Spatial Configuration.* Many of the names used for coordination entities indicate something about the way in which the ligands are bonded to the central atom, something about the geometry of the coordination entity, or both, without explicitly giving all the details. The names in the examples that follow have been formulated using either Stock or Ewens–Bassett numbers, the kappa or hapto convention, and simple methods of designating spatial relationships. Other equally acceptable names can be formulated by the use of variants to these conventions described in this chapter and in Chapter 16.

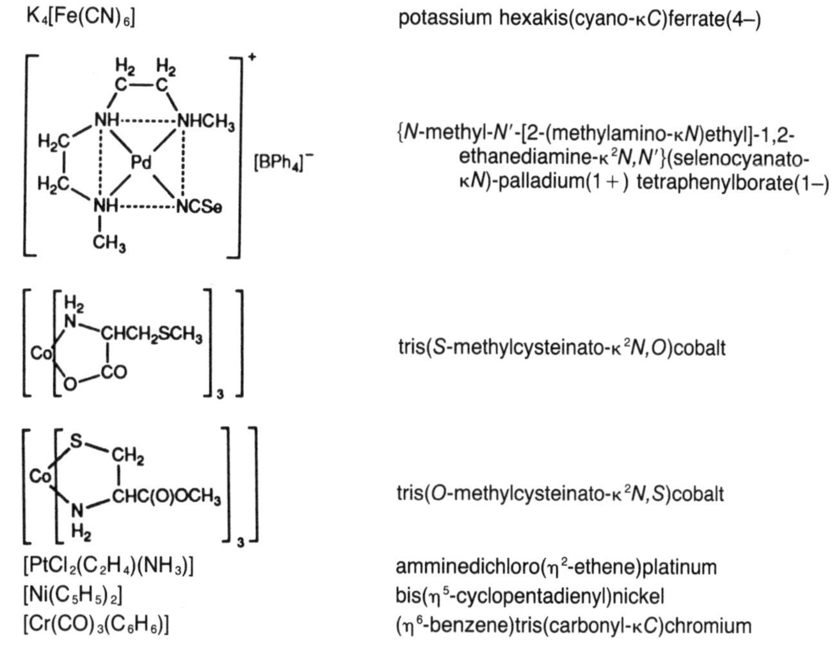

| | |
|---|---|
| $K_4[Fe(CN)_6]$ | potassium hexakis(cyano-κ*C*)ferrate(4–) |
| (structure) $[BPh_4]^-$ | {*N*-methyl-*N'*-[2-(methylamino-κ*N*)ethyl]-1,2-ethanediamine-κ$^2$*N,N'*}(selenocyanato-κ*N*)-palladium(1+) tetraphenylborate(1–) |
| (structure) | tris(*S*-methylcysteinato-κ$^2$*N,O*)cobalt |
| (structure) | tris(*O*-methylcysteinato-κ$^2$*N,S*)cobalt |
| $[PtCl_2(C_2H_4)(NH_3)]$ | amminedichloro(η$^2$-ethene)platinum |
| $[Ni(C_5H_5)_2]$ | bis(η$^5$-cyclopentadienyl)nickel |
| $[Cr(CO)_3(C_6H_6)]$ | (η$^6$-benzene)tris(carbonyl-κ*C*)chromium |

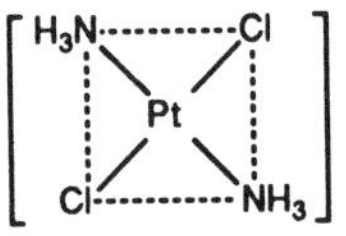

*trans*-diamminedichloroplatinum

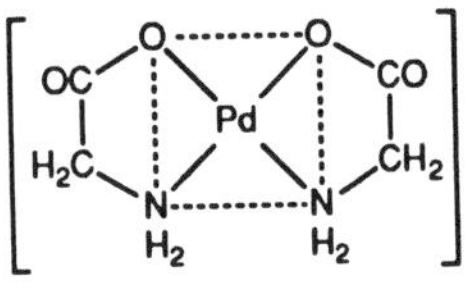

*cis*-bis(glycinato-κ²*N*,*O*)palladium

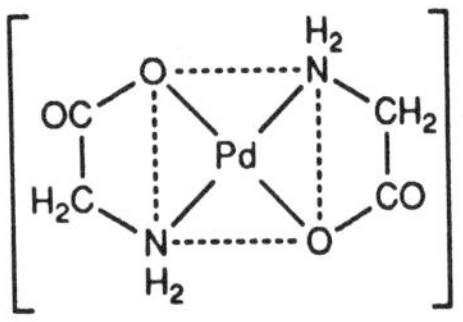

*trans*-bis(glycinato-κ²*N*,*O*)palladium

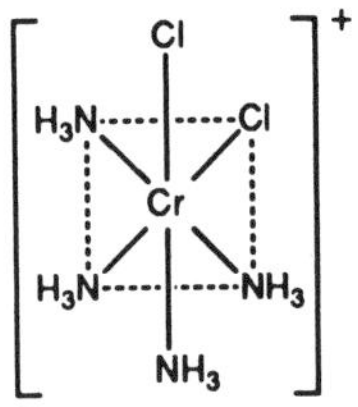

*cis*-tetraamminedichlorochromium(1+) ion

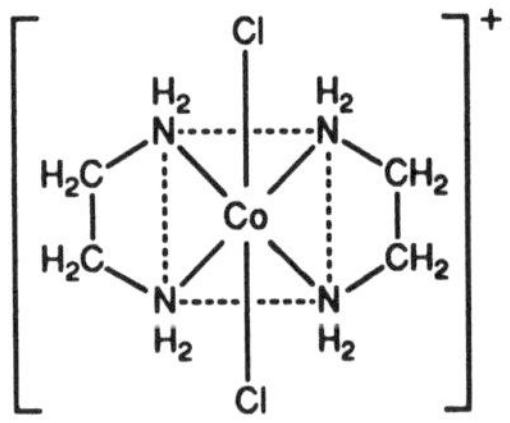

*trans*-dichlorobis(1,2-ethanediamine-κ²*N*,*N*′)-cobalt(1+) ion

## 5.4 Naming Other Inorganic Species

The use of additive nomenclature for inorganic compounds, which is encouraged insofar as possible, requires the selection of a central atom around which to build the name. Additive nomenclature is particularly useful when the central atom is a nonmetal and the species being named is not saltlike in character. Criteria for selecting the central atom are suggested in Section 5.2. The atoms or groups bonded to the central atom will be referred to as side groups in this book, although elsewhere they are sometimes called "ligands" by extension of the use of the term in coordination chemistry (*see also* Section 4.3.1).

The general pattern for additive names is the same as that for names for coordination entities (*see* Section 5.3.1), and the side groups are named in the same manner as ligands (*see* Section 5.3.4). These names for side groups will be referred to as "ligand names" to distinguish them from the "substituent names" used for side groups in substitutive nomenclature (*see* Chapter 9). The "ligand" and "substituent" names for a group are not always the same. The atoms through which side groups bond may be indicated in the same ways as are ligating atoms (*see* Section 5.3.5), and spatial relationships may also be indicated in the same way as they are for coordination entities (*see* Section 5.3.6). There can be more than one additive name for a given species, depending upon the information to be conveyed. For example, $H_2SO_4$ might be named dihydroxodioxosulfur(VI) if viewed as $S(O)_2(OH)_2$ or hydrogen tetraoxosulfate(2–) if viewed as $(H^+)_2(SO_4^{2-})$.

Other systems for naming inorganic species will be discussed in later chapters. These are either older systems that are so embedded in everyday language that they are almost impossible to displace or systems devised for species for which purely additive names are extremely unwieldy. For the most part, such systems are based on modifying the names of parent compounds to specify the particular derivative in question.

In the examples that follow, the names given correspond to the formulas as written and give no further information.

| | |
|---|---|
| $Se(O)_2(F)_2$ | difluorodioxoselenium(VI) |
| $P(S)(Cl)_3$ | trichlorosulfidophosphorus(V) |
| $S(O)_3(F)^-$ | fluorotrioxosulfate(1–) ion |
| $PCl_4^+$ | tetrachlorophosphorus(1+) ion |
| $F_5SOS(O)_2OSF_5$ | dioxobis[pentafluorooxosulfato(1–)]sulfur(VI) |
| $FS(O)_2OP(O)(F)_2$ | difluoro[fluorotrioxosulfato(1–)]oxophosphorus(V) |
| F, Cl, Cl, P, Cl, F (structure) | trichloro-*trans*-difluorophosphorus |

# Chapter 6
# Polynuclear Coordination Entities

## 6.1 Introduction

Although most coordination compounds contain only a single center of coordination, a large number contain more than one. These polynuclear species, which may be neutral or ionic, exist in a variety of structural forms; both metallic and nonmetallic elements may serve as multiple centers in chains and rings, polymers, clusters, and extended assemblies of oxoanions. The linkage between centers may be through a bridging group, or the centers may be directly linked to each other. This chapter is primarily concerned with finite structures in which there are two or more coordination centers. The following chapter, which is devoted to inorganic polymers, includes polymeric species containing coordination centers in the repeating unit.

The general principles for naming coordination compounds (*see* Section 5.3) also apply to polynuclear coordination entities. In compositional names the bridging groups are treated as ligands and included in proper alphabetical order with the nonbridging ligands, and the central atoms are listed in alphabetical order after the ligands. A ligand is cited first as a bridging group, which is indicated by the symbol μ (*see* Section 6.2), and then as a nonbridging ligand. Multiple bridging is cited in decreasing order of complexity; that is, a ligand bridging three centers is cited before the same ligand bridging two centers, etc. The number of each kind of central atom is indicated by a type-1 numerical prefix (*see* Table A.II).

Examples of compositional names:

| | |
|---|---|
| $[(OC)_3Fe(CO)_3Fe(CO)_3]$ | tri-μ-carbonyl-hexacarbonyldiiron |
| $[(C_8H_{12})Rh(Cl)_2Rh(C_8H_{12})]$ | di-μ-chloro-bis(η$^4$-1,5-cyclooctadiene)dirhodium |
| $[(H_3N)_4Co(OH)_2Co(NH_3)_4]Cl_4$ | octaammine-di-μ-hydroxo-dicobalt(4+) chloride |

More informative names for species containing bridged centers of coordination are derived using the principles presented in the following sections of this chapter.

## 6.2 Designation of Bridging Groups

Bridging groups may be single atoms or polyatomic groups. Polyatomic groups may bridge through one atom of the group or through two or more atoms of the group. Bridging groups are indicated by placing the Greek letter μ, separated by a hyphen, before the ligand name of the group and setting off the ligand name with hyphens. If the ligand name requires enclosing marks, μ is placed inside the enclosing marks. When it is desired to indicate the specific ligating atom or atoms of the bridging group, italicized symbols for the elements involved may be placed after the name for the bridging group (*see* Section 5.3.5.1), and the kappa convention (*see* Section 5.3.5.2) may be used. If the bridging takes place via different atoms of the same group, the ligating locants and symbols indicating the bonding to different central atoms are separated by a colon. The number of coordination centers bridged by a bridging group is indicated by a subscript following μ (i.e., $\mu_n$, where $n$ is the number of coordination centers linked together by the bridging group). When the number of coordination centers bridged is two ($n = 2$), the subscript is usually omitted. The examples in the remainder of this chapter illustrate how bridging groups are named.

## 6.3 Dinuclear Species with Bridging Groups

### 6.3.1 Symmetrical Species

Many dinuclear coordination entities involve two identical units held together by one or more bridging groups. Simple names for such species may often be formed according to principles of multiplicative nomenclature by citing multiples of these identical units. Such names begin with citation of the bridging group or groups, followed by the name of the nuclear system with the type-2 numerical prefix, "bis" (*see* Table A.II). Ewens–Bassett numbers refer to the charge on the entire unit, whereas Stock numbers refer to the oxidation numbers of individual central atoms.

Examples:

| | |
|---|---|
| $[(H_3N)_5Cr\text{-}OH\text{-}Cr(NH_3)_5]^{5+}$ | μ-hydroxo-bis(pentaammine-chromium)(5+) ion |
| $[(EtPh_2P)_3(N_2)_2W\text{-}N_2\text{-}W(PEtPh_2)_3(N_2)_2]$ | μ-dinitrogen-κ*N*:κ*N*′-bis[bis(dinitrogen)-tris(ethyldiphenylphosphine)tungsten] |
| $[(NCSSCN)(Cl)_2(O)_2Cr\text{-}NCSSCN\text{-}Cr(O)_2(Cl)_2(NCSSCN)]$ | μ-thiocyanogen-κ*N*:κ*N*′-bis[dichlorodioxo-(thiocyanogen-κ*N*)chromium(VI)] |
| $[Br_2Pt(\text{-}SMe_2\text{-})_2PtBr_2]$ | bis-(μ-dimethyl sulfide-κ²*S*)-bis[dibromo-platinum(II)] |
| $[(OC)_3Re(\text{-}I\text{-})_3Re(CO)_3]$ | tri-μ-iodo-bis(tricarbonylrhenium) |
| $[(H_3N)_3Co(\text{-}OH\text{-})_2(\text{-}NO_2\text{-})Co(NH_3)_3]Br_3$ | di-μ-hydroxo-μ-nitro-κ*N*:κ*O*-bis(triammine-cobalt)(3+) bromide |

### 6.3.2 Unsymmetrical Species

Bridged dinuclear entities may be unsymmetrical because different central atoms are present or because the ligands bonded to identical central atoms are different. They are named in all cases by citing the ligands and bridging groups in a single alphabetically ordered series, followed by the names of the central atoms in alphabetical order. The central atoms are numbered 1 and 2 to provide numerical locants for the ligands and locants for the points of attachment of bridging groups.

The CNIC recommendations for numbering the central atoms follow (*30*). If the central atoms are different, the higher priority central atom (i.e., the atom to be numbered 1) is the more electropositive element according to Figure A.1. If the central atoms are identical, the higher priority central atom is the central atom that gives the lower locant set for ligands at the first point of difference. Should that procedure fail, the alphabetical order of ligands is used to establish priority (i.e., the central atom with the larger number of ligands with initial letters earlier in the alphabet is numbered 1). The order of citation and the order of numbering may not be the same when the central atoms are different. These criteria for numbering identical central atoms seem rather complicated, and the following criteria from Section 5.2 for selecting the atom to be numbered 1, with priority decreasing from 1 to 4, might be simpler.

1. Atom with higher oxidation number
2. Atom with higher coordination number
3. Atom with greater variety of bonding atoms
4. Atom to which is attached the side group with the name first in alphabetical order

The kappa convention (*see* Section 5.3.5.2) is used in modified form to indicate the ligating atom(s) and its (their) distribution. The use of a right superscript numeral on κ to indicate the number of equivalent ligating atoms bonded to a central atom is retained. In addition, the numerical locant of the central atom to which the ligand is bonded is placed on the line before the κ.

Examples:

|  |  |
|---|---|
| $[(H_3N)_5\underset{1}{\mathrm{Ru}}\text{-NCC}(C_2H_4)_3\text{CCN-}\underset{2}{\mathrm{Co}}(NH_3)_3(OH_2)_2]^{6+}$ | octaammine-1κ$^{5}$*N*,2κ$^{3}$*N*-diaqua-2κ$^{2}$*O*-(μ-bicyclo[2.2.2]octane-1,4-dicarbonitrile-1κ*N*:2κ*N′*)-2-cobalt-1-ruthenium(6+) ion |
| $[(Et_2PhP)_3(Cl)_2\underset{1}{\mathrm{Re}}\text{-N-}\underset{2}{\mathrm{Pt}}(PEt_3)(Cl)_2]$ | tetrachloro-1κ$^{2}$,2κ$^{2}$-tris(diethylphenylphosphine-1κ*P*)-μ-nitrido-(triethylphosphine-2κ*P*)-2-platinum-1-rhenium |
| $[(H_3N)_5\underset{1}{\mathrm{Cr}}(\text{-OH-})\underset{2}{\mathrm{Cr}}(NH_3)_4(NH_2Me)]Cl_5$ | nonaammine-1κ$^{5}$*N*,2κ$^{4}$*N*-μ-hydroxo(methanamine-2κ*N*)dichromium(5+) chloride |
| $[(H_3N)_4(Cl)\underset{1}{\mathrm{Ni}}\text{-SC(Me)O-}\underset{2}{\mathrm{Ni}}(NH_3)_3(Cl)_2]$ | heptaammine-1κ$^{4}$*N*,2κ$^{3}$*N*-trichloro-1κ,2κ$^{2}$-(μ-thioacetato-2κ*O*:1κ*S*)-dinickel |
| $[(H_3N)_4(Cl)\underset{1}{\mathrm{Ni}}\text{-OC(Me)S-}\underset{2}{\mathrm{Ni}}(NH_3)_3(Cl)_2]$ | heptaammine-1κ$^{4}$*N*,2κ$^{3}$*N*-trichloro-1κ,2κ$^{2}$-(μ-thioacetato-1κ*O*:2κ*S*)-dinickel |

## 6.4 Other Polynuclear Species with Bridging Groups

### 6.4.1 General

The nomenclature for bridged polynuclear species more complex than binuclear is based on a logical procedure for numbering the atoms of the central or fundamental structural unit (CSU). Only the central atoms are considered for this purpose. Descriptors such as *cyclo-*, *tetrahedro-*, *octahedro-*, etc. (*see* Table A.III) have traditionally been used to describe nonlinear CSUs. Established structures for CSUs now exceed the limited set of descriptors associated with this usage. Their number makes it necessary to supplement or replace this set with a more comprehensive system. The so-called "CEP system" developed by Casey, Evans, and Powell (*53*) for triangulated polyboron polyhedrons (deltahedrons) can be adapted to provide systematic descriptors for the geometrical figures formed by the metal atoms (i.e., if the bridging groups are ignored). Some of these descriptors are listed in Table A.VI. The descriptor for the CSU, with all Roman letters italicized, should appear in the name just before the central atoms are cited. The scheme used for numbering polyhedral CSUs involves the selection of a reference axis (an axis of highest rotational symmetry) and the assignment of atoms to planes perpendicular to the reference axis. The procedure is similar to that used for assigning locants to the bonding positions of a central atom (*see* Section 16.2.2).

Central atoms in chain and branched-chain polynuclear structures are numbered consecutively, proceeding from one end along the path containing the greatest number of central atoms. The end of a chain from which to start numbering is the end with the more electropositive central atom (*see* Figure A.1). If the end central atoms are the same, numbering starts from the end that gives the lower locant to the most electropositive central atom within the chain. If there is still a choice for numbering, preference is given to the lower locant set for the ligands according to Section 2.2.4.6. Central atoms in monocyclic polynuclear structures are numbered consecutively, starting with the most electropositive central atom in the ring and proceeding in the direction of the nearest central atom of the same kind or, as a next choice, in the direction of the nearest next most electropositive central atom in the ring. If a choice in numbering still remains, the central atoms are numbered to give the lower locant set of numbers for ligands (*see* Section 2.2.4.6).

In all cases the locant for each central atom, set off by hyphens, is placed before that atom's name in the central atom list when there is any possibility of ambiguity. The kappa convention as modified in Section 6.3.2 is used as necessary to indicate the position of ligands and bridging groups and the kinds and numbers of ligating atoms in each ligand or bridging group.

### 6.4.2 Symmetrical Species

Symmetrical structures for bridged trinuclear and higher species may result from the association of identical units with one another or with another group or groups. Such species can be named unambiguously without including all of the locants provided for in Section 6.4.1 because there is no possibility for isomerism. In many cases multiplicative nomenclature can be used with type-2 numerical prefixes (*see* Table A.II) in conjunction with names of identical units to obtain simpler names. A Ewens–Bassett number, which refers to the charge on the entire entity, is placed outside the multiplicative enclosing marks, whereas Stock numbers, which refer to the oxidation numbers of individual central atoms, are placed inside them.

Examples:

| | |
|---|---|
| $[(MeHg)_4S]^{2+}$ | $\mu_4$-thio-tetrakis(methylmercury)(2+) ion |
| $[Be_4O(OAc)_6]$ | hexakis-($\mu$-acetato-κ*O*:κ*O'*)-$\mu_4$-oxo-*tetrahedro*-tetraberyllium |
| | hexakis-($\mu$-acetato-κ*O*:κ*O'*)-$\mu_4$-oxo-[$T_d$-(13)-$\Delta^4$-*closo*]tetraberyllium |
| $[Mo_6S_8]^{2-}$ | octa-$\mu_3$-thio-*octahedro*-hexamolybdate(2–) |
| | octa-$\mu_3$-thio-[$O_h$-(141)-$\Delta^8$-*closo*]hexamolybdate(2–) |
| $[(Me_3PtI)_4]$ | tetra-$\mu_3$-iodo-tetrakis[trimethylplatinum(IV)] |
| | tetra-$\mu_3$-iodo-dodecamethyl-1κ$^3$*C*,2κ$^3$*C*,3κ$^3$*C*,4κ$^3$*C*-*tetrahedro*-tetraplatinum |
| | tetra-$\mu_3$-iodo-dodecamethyl-1κ$^3$*C*,2κ$^3$*C*,3κ$^3$*C*,4κ$^3$*C*-[$T_d$-(13)-$\Delta^4$-*closo*]tetraplatinum |

### 6.4.3 Unsymmetrical Species

If the name for a species cannot be simplified by considerations of symmetry, the name is derived by using the general principles set forth in Section 6.4.1. It is conventional in applying the kappa convention to count the number of contacts that a bridging group makes with central atoms; that is, the bridging group in ...Cr(OH)Co... would be named "μ-hydroxo-1:2κ$^2$*O*", the bridging groups in ...$Cr(OH)_2Co$... would be named "di-μ-hydroxo-1:2κ$^4$*O*", etc. When it is necessary to distinguish among ligating atoms, locants based on the ligand are added as superscripts to the right of the italicized symbol for the ligating atom.

Examples:

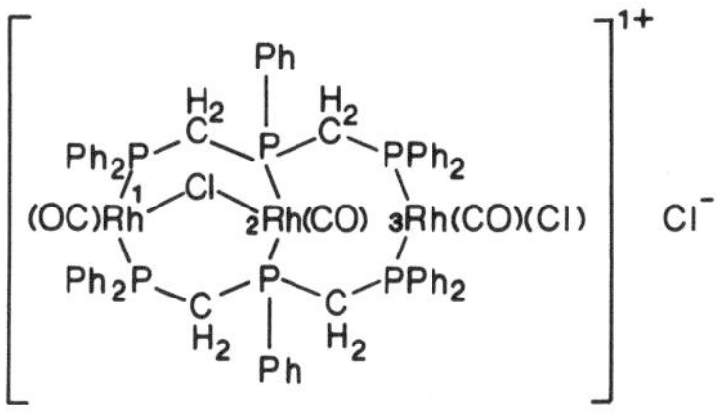

tricarbonyl-1κ*C*,2κ*C*,3κ*C*-μ-chloro-1:2κ$^2$-chloro-3κ-bis-{μ$_3$-bis[(diphenylphosphino)-1κ*P*:3κ*P*′-methyl]phenylphosphine-2κ*P*}-trirhodium(1+) chloride

Comment: The rhodium atoms are numbered to give the bridging chloro the locant 1:2κ$^2$ instead of the higher 2:3κ$^2$; this is the first point of difference.

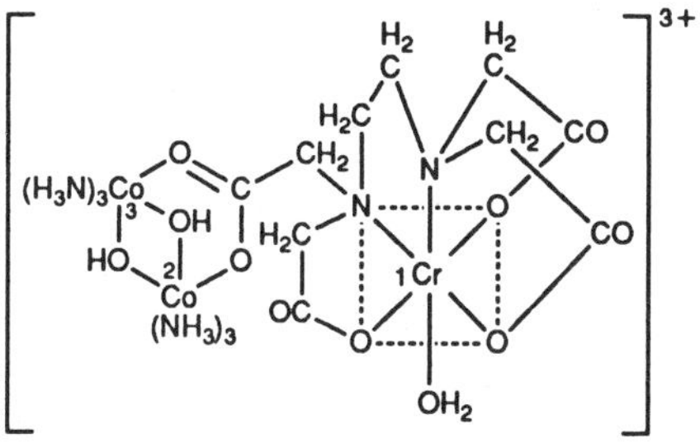

hexaammine-2κ$^3$*N*,3κ$^3$*N*-aqua-1κ*O*-[μ$_3$-1,2-ethanediyldinitrilo-1κ$^2$*N*,*N*′-tetraacetato(4−)-1κ$^3$*O*$^1$,*O*$^{1''}$,*O*$^2$:2κ*O*$^{2''}$:3κ*O*$^{2'''}$]-di-μ-hydroxo-2:3κ$^4$*O*-1-chromium-2,3-dicobalt(3+) ion

Comments: The more electropositive chromium atom is numbered 1, and the cobalt atoms are numbered 2 and 3. Priming is used to distinguish among the carboxylate oxygen atoms. In accordance with CAS index nomenclature, the oxygen atoms of one acetato group are designated *O*$^1$(⩾C—O—) and *O*$^{1'}$ (>C=O), those of the other acetato group attached to the same nitrogen atom *O*$^{1''}$ (⩾C—O—) and *O*$^{1'''}$ (>C=O). The oxygen atoms of the acetato groups attached to the second nitrogen atom are similarly designated *O*$^2$, *O*$^{2'}$, *O*$^{2''}$, and *O*$^{2'''}$. The ligating atoms bonding the three acetato groups to the chromium atom are designated *O*$^1$, *O*$^{1'}$, and *O*$^2$ (the lowest locant set). The cobalt atom bonded to O$^{2''}$ is numbered 2 because *O*$^{2''}$ is lower than *O*$^{2'''}$. In the 1990 edition of the IUPAC Inorganic Rules (*30*), the ligating oxygen atoms are designated *O*$^1$, *O*$^2$, *O*$^3$, *O*$^4$, and *O*$^{4'}$.

The central atoms in cyclic polynuclear species are assigned priority according to the principles in Section 6.4.1, and the italicized prefix *cyclo-* is used at the beginning of the name to indicate a monocyclic species.

Examples:

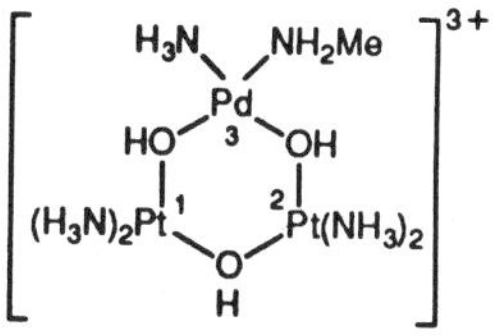

*cyclo*-pentaammine-1κ$^2$*N*,2κ$^2$*N*,3κ*N*-tri-μ-hydroxo-1:2κ$^2$*O*;1:3κ$^2$*O*;2:3κ$^2$*O*-(methanamine-3κ*N*)-3-palladium-1,2-diplatinum(3+) ion

Comment: The more electropositive platinum atoms, which are equivalent, are numbered 1 and 2, and the palladium atom is numbered 3.

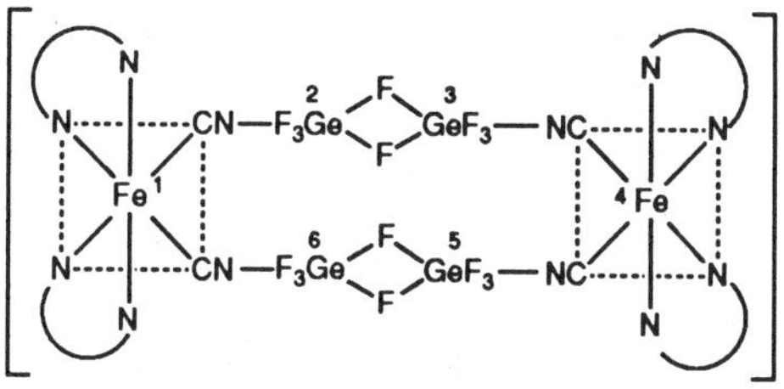

*cyclo*-tetra-μ-cyano-1κ*C*:2κ*N*; 1κ*C*:6κ*N*; 3κ*N*:4κ*C*;-4κ*C*:5κ*N*-tetrakis(4,7-diphenyl-1,10-phenanthroline)-1κ$^2$*N*,*N*′; 1κ$^2$*N*,*N*′; 4κ$^2$*N*,*N*′;-4κ$^2$*N*,*N*′-tetra-μ-fluoro-2:3κ$^2$*F*; 2:3κ$^2$*F*; 5:6κ$^2$*F*;-5:6κ$^2$*F*-dodecafluoro-2κ$^3$,3κ$^3$,5κ$^3$,6κ$^3$-2,3,5,6-tetragermanium-1,4-diiron

Comment: (N N represents 4,7-diphenyl-1,10-phenanthroline

## 6.5 Polynuclear Species without Bridging Groups

### 6.5.1 General

A number of compounds contain metal–metal bonds. Metal–metal bonding is indicated by placing the italicized symbols of the appropriate atoms, separated by a long dash and enclosed in parentheses, after the list of central atoms, but before the Ewens–Bassett number. This symbolism is not intended to specify that the metal–metal bond is a simple covalent bond; it merely indicates connectivity. The number of metal–metal bonds of one kind (i.e., the pair of atoms in each bond is the same) in the species being named is indicated by placing the appropriate arabic numeral inside the parentheses, in front of the atomic symbols, and separated from them by a space. If there is more than one kind of metal–metal bond in a given species, each kind is indicated with a separate parenthetical expression. Such parenthetical expressions are listed in alphabetical order of the symbols.

Assemblies of more than three atoms with direct linkages to form three-dimensional structures are generally termed "cluster compounds". The nomenclature of polynuclear species with metal–metal bonds is fundamentally the same as that for polynuclear species with bridging groups between coordination centers; that is, the ligands are cited in one alphabetical series, followed by citation of the coordination centers in a second alphabetical series. Appropriate locants and descriptors are inserted where needed to complete the name (*see* Section 6.4.1).

### 6.5.2 Symmetrical Species

Multiplicative nomenclature employing type-2 numerical prefixes (*see* Table A.II) is used whenever feasible to form names as simple as possible.

Examples:

| | |
|---|---|
| $[Br_4ReReBr_4]^{2-}$ | bis(tetrabromorhenate)(*Re–Re*)(2–) ion |
| $[(OC)_5MnMn(CO)_5]$ | bis(pentacarbonylmanganese)(*Mn–Mn*) |
| $[(PhNC)_3PdPd(CNPh)_3]$ | bis[tris(phenyl isocyanide)palladium](*Pd–Pd*) |
| $[\{Os(CO)_4\}_3]$ | *cyclo*-tris(tetracarbonylosmium)(3 *Os–Os*) |
| $[Os(CO)_4\{Os(CO)_4(SiCl_3)\}_2]$ | tetracarbonylbis[tetracarbonyl(trichlorosilyl)-osmium]osmium(2 *Os–Os*) |
| $Cs_3[(ReCl_4)_3]$ | cesium *cyclo*-tris(tetrachlororhenate)(3 *Re–Re*)(3–) |

### 6.5.3 Unsymmetrical Species

Priorities and locants are determined in the same way as for polynuclear species with bridging groups between coordination centers (*see* Section 6.4.1).

Examples:

| | |
|---|---|
| $[(OC)_4CoRe(CO)_5]$ (Co = 2, Re = 1) | nonacarbonyl-1κ$^5$*C*,2κ$^4$*C*-2-cobalt-1-rhenium(*Co–Re*) |
| $[(Ph_3As)AuMn(CO)_5]$ (Au = 2, Mn = 1) | pentacarbonyl-1κ$^5$*C*-(triphenylarsine-2κ*As*)-2-gold-1-manganese(*Au–Mn*) |
| $[(Ph_3P)_2(OC)(Cl)_2IrHgCl]$ (Ir = 1, Hg = 2) | carbonyl-1κ*C*-trichloro-1κ$^2$,2κ-bis(triphenylphosphine-1κ*P*)-1-iridium-2-mercury(*Ir–Hg*) |
| $[(Ph_3P)_2Pt\{Fe(CO)_4\}_2]$ (Fe = 1, 2; Pt = 3; triangle Pt–Fe–Fe) | octacarbonyl-1κ$^4$*C*,2κ$^4$*C*-bis(triphenylphosphine-3κ*P*)-*triangulo*-1,2-diiron-3-platinum(*Fe–Fe*)-(2 *Fe–Pt*) |

Comment: The descriptor *cyclo*- could be used instead of *triangulo*-.

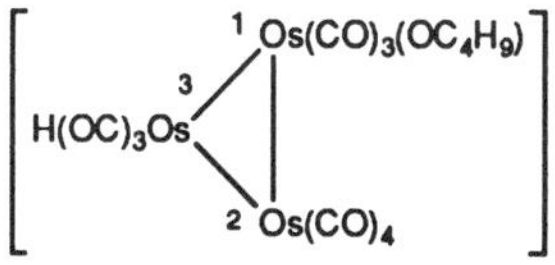

$^{1}Os(CO)_3(OC_4H_9)$ $^{3}$ $H(OC)_3Os$ $^{2}Os(CO)_4$

butoxo-1κ*O*-decacarbonyl-1κ$^3$*C*,2κ$^4$*C*,3κ$^3$*C*-hydrido-3κ-*triangulo*-triosmium(3 *Os–Os*)

Comments: Numbering the Os atoms as shown gives the lowest locants to the ligands that are earlier alphabetically. That is, 1κ*O* is lower than 2κ*O* or 3κ*O* for butoxo, and 1κ$^3$2κ$^4$3κ$^3$ is lower than 1κ$^3$2κ$^3$3κ$^4$ for decacarbonyl. Using the criteria suggested in Section 6.3.2 would reverse 2 and 3.

## 6.6 Species with Metal–Metal Bonds and Bridging Groups

The naming of polynuclear coordination entities with both metal–metal bonds and bridging groups poses no additional problems. Such species are named as bridged compounds, and the metal–metal bonds are indicated as in Section 6.5. The indication of metal–metal bonds has frequently been omitted, particularly for multiatom homoatomic aggregates.

Examples:

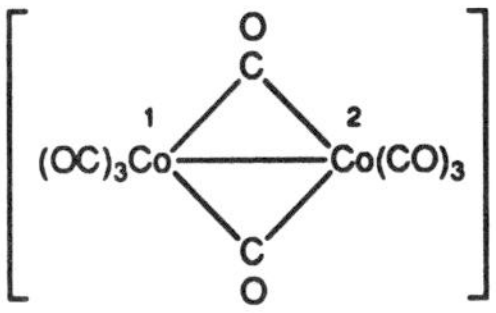

O C $(OC)_3Co$ $^1$ $^2$ $Co(CO)_3$ C O

bis-(μ-carbonyl-1:2κ$^2$*C*)-bis(tricarbonyl-κ$^3$C-cobalt)(*Co–Co*)

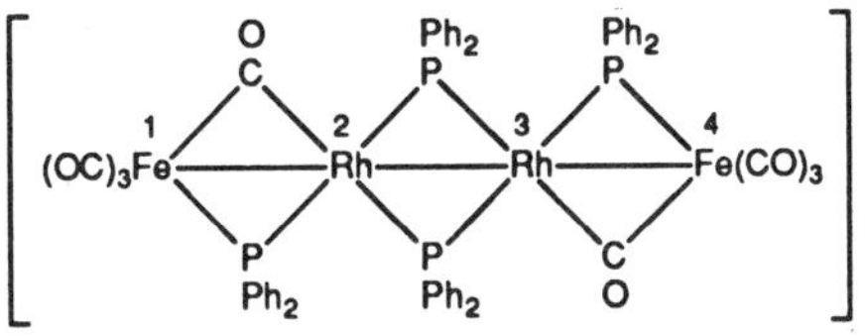

O C $Ph_2$ P $Ph_2$ P $(OC)_3Fe$ $^1$ $^2$ Rh $^3$ Rh $^4$ $Fe(CO)_3$ P $Ph_2$ P $Ph_2$ C O

di-μ-carbonyl-1:2κ$^2$*C*;3:4κ$^2$*C*-hexacarbonyl-1κ$^3$*C*,4κ$^3$*C*-tetrakis-(μ-diphenylphosphido)-1:2κ$^2$*P*;2:3κ$^4$*P*;3:4κ$^2$*P*-1,4-diiron-2,3-dirhodium(2 *Fe–Rh*)(*Rh–Rh*)

Comment: The chain of metal atoms may be numbered starting from either end.

$[(Mo_6Cl_8)Cl_6]^{2-}$ octa-μ$_3$-chloro-hexachloro-*octahedro*-hexamolybdate(12 *Mo–Mo*)(2–) ion

$[Nb_6Cl_{12}]^{2+}$ dodeca-μ-chloro-*octahedro*-hexaniobium-(12 *Nb–Nb*)(2+) ion

Comment: The descriptor (12 *Nb–Nb*) only indicates that there are 12 *Nb–Nb* connections, not that there are 12 discrete covalent metal–metal bonds, and it may be omitted.

# Chapter 7
# Polymeric Inorganic Species

## 7.1 Introduction

Each polymeric substance usually includes a number of different individual polymeric molecules and/or ions, and a complete description would have to include such items as degree of steric regularity, chain imperfections, random branching, etc., for each different polymeric molecule in the substance. Such a description would result in an extremely complex nomenclature, so it is useful to consider a polymer in terms of a single idealized structure. Prior to the 1980s there was no general method for naming inorganic polymers. Different kinds of inorganic polymers were named either according to generally accepted practices for organic polymers, with allowances for the lack of names for inorganic groups corresponding to organic "biradical" groups, or by an extension of coordination nomenclature.

The system of nomenclature devised for regular single-strand and quasi-single-strand inorganic and coordination polymers (*28*) is based on selection and naming of a preferred constitutional repeating unit to provide a unique name. Principles of inorganic and coordination nomenclature are used insofar as they are consistent with principles of polymer nomenclature. Because differences between the methods for naming polynuclear coordination entities and for naming polymers led to different names, it was necessary to make a choice. The decision to follow the established polymer conventions gives names that differ from those derived by following the inorganic practice for citing bridging groups (*see* Section 6.1).

It is useful to introduce some terminology of polymer nomenclature at this point. The constitutional repeating unit (CRU) is the smallest unit of structure whose repetition describes the polymer structure. A regular single-strand polymer is a linear polymer that can be described by a preferred CRU in which both terminal constituent subunits of the CRU are connected through single atoms to the other CRUs or to end groups (i.e., ...$\leftarrow$A{• • • • •}B$\rightarrow_n$...). A quasi-single-strand polymer is a linear polymer that can be described by a preferred CRU in which only one terminal constituent subunit is connected through a single atom to the other CRUs or to an end group (i.e., ...$\leftarrow$A{• • • • •}B$\rightarrow_n$...). Inorganic and coordination quasi-single-strand polymers, although not fitting the definition of a regular single-strand polymer, can be named in the same manner as such a polymer is named.

The name of a polymer for which the CRU is known, but with a dimensional structure that is unknown or need not be specified, consists of the prefix "poly" followed by the name of the CRU enclosed in square brackets (i.e., poly[CRU]). The number of CRUs in a polymer structure may be specified by an appropriate numerical prefix in place of the prefix poly- (example: deca[CRU]). A linear (i.e., one-dimensional) polymer is indicated by the italicized prefix "*catena*-" added to the name of the polymer. Procedures for naming two- and three-dimensional polymers have not yet been agreed upon, but it has been suggested that the two types could be indicated by the italicized prefixes "*phyllo*-" and "*tekto*-"*, respectively (*28*).

End groups of a polymer molecule are specified, if desired, by appropriate names, which are identified by the Greek letters α and ω and are added to the name of the polymer as hyphenated prefixes {example: α-(end group)-ω-(end group)-*catena*-poly[CRU]}.

## 7.2 Constitutional Repeating Unit

### 7.2.1 Identification

In many cases the polymer structure is simple enough that the preferred CRU and its constituent subunits can be identified readily. For example, in the polymer ...Au(F)FAu(F)F..., the two possible CRUs are –Au(F)F– and –FAu(F)–. However, in more complex cases, and occasionally in some simple cases, it may be necessary to draw a fairly long segment of the polymer chain in order to identify the possible CRUs.

In the polymer ...AgCNAgCN..., the possible CRUs are –AgCN–, –AgNC–, –CNAg–, and –NCAg– if the constitutional subunits are Ag and CN. There are four possible CRUs in the polymer ...$HgSC(NEt_2)OHg(Br)_2OC(NEt_2)SHgOC(NEt_2)SHg(Br)_2SC(NEt_2)O$... if the constitutional subunits are Hg, $SC(NEt_2)O$, and $HgBr_2$.

The possible constitutional subunits of a CRU in regular single-strand and quasi-single-strand inorganic and coordination polymers are single central atoms, central atoms with associated side groups, and bridging groups. Thus, the subunits of the CRUs in the preceding paragraph are, in fact, –Ag– and μ-CN– or μ-NC– in the first example and –Hg–, $-HgBr_2-$, and $\mu$-$SC(NEt_2)O$– or $\mu$-$OC(NEt_2)S$– in the second. All inorganic and coordination polymers have one or more central atoms, but they may or may not have bridging groups.

Selection of the largest structural fragments in the backbone that can be assigned multivalent substituent names as subunits of a CRU is a fundamental principle in naming linear organic polymers. In naming inorganic and coordination polymers, this principle is applied to the selection of bridging groups. When there is a choice, the largest group that can be named by the accepted methods for naming polydentate groups is chosen. For example, in the polymer

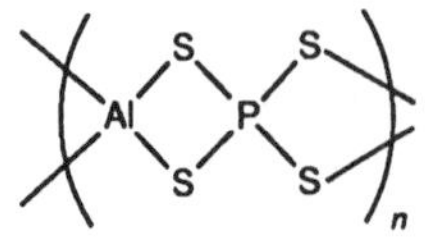

*The spelling "*tekto*-" is derived from the Greek τεκτον-, whereas "*tecto*-" (*221*) is derived from the Latin *tecton*-.

the CRU could be considered to consist of two central atoms connected by bridging sulfur atoms. The principle of "largest bridging group", however, requires that $[S_2PS_2]^{3-}$ be chosen as the bridging group. Strict application of the principle of largest subunit to inorganic and coordination polymers would lead to the selection of polynuclear coordination centers as the largest fragments in the CRU. Because at present there are no officially adopted procedures for uniquely naming or numbering some polynuclear coordination centers, such species are used as subunits of CRUs only when it is not convenient to express such structural units in terms of their mononuclear coordination centers. An example of a polymer with a CRU containing a polynuclear constituent subunit is

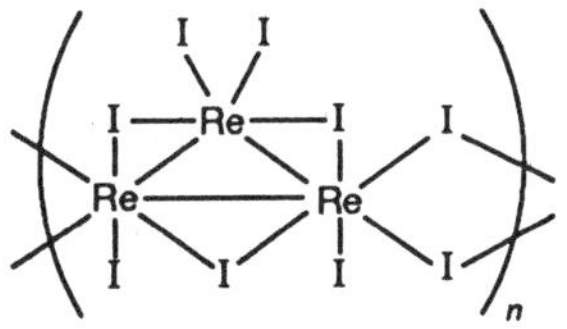

in which the constituent subunits are $Re_3I_7$ and μ-I. The examples given in the remainder of this chapter will further illustrate the selection of constituent subunits in the CRU.

### 7.2.2 Seniority

The constituent subunit of the CRU with the highest seniority is the central atom or coordination center preferred according to the following criteria, applied in order until no further decision is needed. This subunit is written as the left terminal subunit of the CRU.

1. The constituent subunit of highest seniority in the preferred CRU of an inorganic or coordination polymer must contain one or more central atoms (i.e., bridging groups between central atoms in the backbone of the polymer cannot be senior subunits).
2. When there are two or more central atoms in the CRU, the senior subunit is the one containing the most electropositive central atom according to Figure A.1.
3. If the use of polynuclear centers is necessary, the center with the largest number of central atoms is preferred.
4. The central atom or coordination center with the greatest number of directly attached atoms exclusive of atoms from bridging groups in the backbone is preferred.
5. The central atom or coordination center with the name that is earliest alphabetically is preferred. This name includes any side groups and their multiplying prefixes, other than the bridging groups in the backbone of the polymer chain.

### 7.2.3 Orientation

The constituent subunits in the CRU are cited sequentially, starting with the senior subunit. The preferred direction is determined by using the following criteria, which are considered in order until no further decision is needed.

1. A single-strand CRU is preferred to a quasi-single-strand CRU (i.e., a CRU with both terminal constituent subunits connected to other identical CRUs or to an end group through single atoms is preferred to a CRU with only one terminal constituent

subunit connected to other CRUs or to an end group through a single atom). For example,

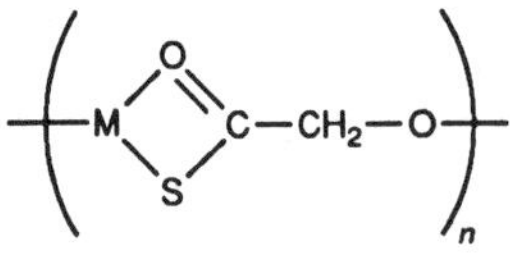

is preferred to

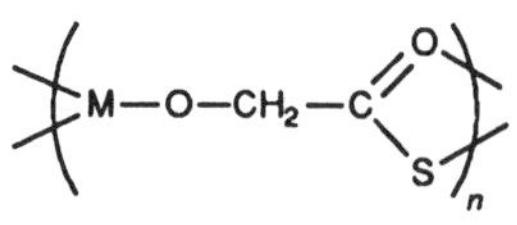

2. The preferred direction is defined by the shortest path, measured in terms of the number of atoms in the most direct continuous chain of atoms in the polymer backbone, from the senior subunit to a subunit of equal seniority or to a subunit next in seniority. For example, $-MS(R)M'NH_2NH_2-$ is preferred to $-MNH_2NH_2M'S(R)-$ because the one-atom path through the thiolato group is preferred to the two-atom path through the hydrazine group, and

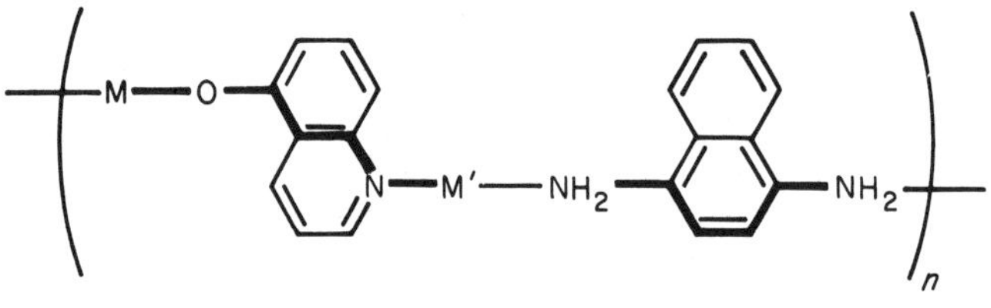

is preferred to

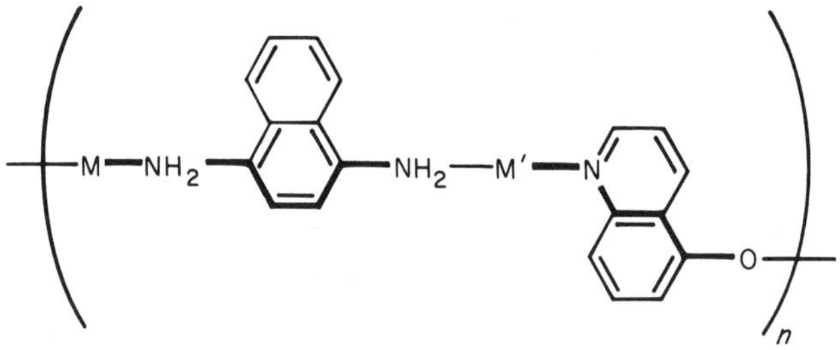

because the five-atom path through the 5-quinolinolato group is preferred to the six-atom path through the 1,4-naphthalenediamine group.

3. When all paths between the senior subunit and a subunit of equal seniority, or a subunit ranking next in seniority, are of equal length, the preferred direction is along the path that includes constituent subunits of higher seniority. The paths between subunits of equal seniority or between the senior subunit and the subunit next in seniority necessarily involve subunits of lesser seniority. When the subunits are organic groups, the hierarchical order of subunits prescribed for linear organic polymers (*54*) may be needed to determine the preferred direction. For example,

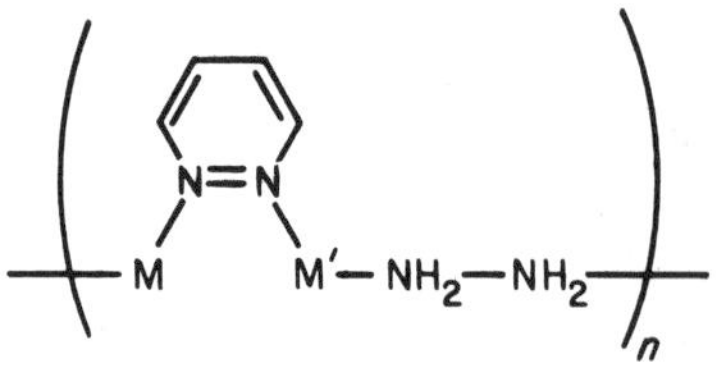

is preferred to

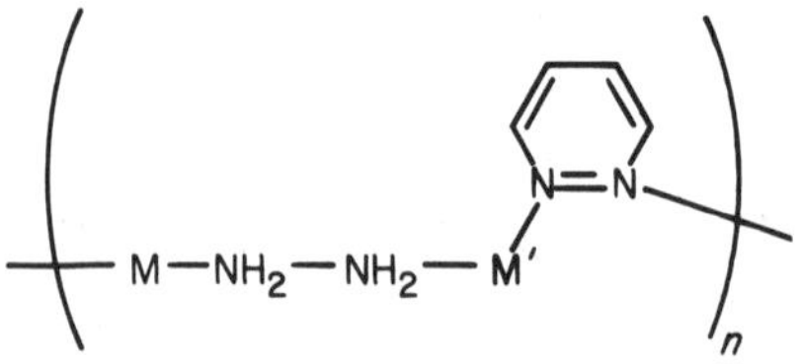

because the heterocyclic structure is preferred to the acyclic hetero chain. The CRU $-M-Cl-M'-OH-$ is preferred to $-M-OH-M'-Cl-$ and $-M-N{\equiv}C-$ is preferred to $-M-C{\equiv}N-$. In each case, the preference is based on organic heteroatom priority.

4. If a further choice is needed, the preferred path is the one containing the group that has the name coming first alphabetically at the first difference.

### 7.2.4 Name

Names of CRUs of single-strand and quasi-single-strand inorganic and coordination polymers are based on a backbone consisting of central atoms and, where present, bridging groups. Coordination centers, mononuclear or polynuclear, and their associated groups (except for groups bridging between central atoms in the backbone, if any) are named by the system used for coordination entities (*see* Chapters 5 and 6). Bridging groups are named by the system used for bridging ligands in polynuclear species (*see* Section 6.2), except that the μ designating a bridging ligand in the backbone of a polymer is not placed inside the enclosing marks.

Italicized element symbols indicating the atoms of bridging groups linking the backbone are cited in the order of direction of decreasing seniority in the CRU and are separated by colons. Hence, element symbols cited before a colon refer to bonds to the central atom occurring just before the bridging group in the CRU, whereas element symbols cited after a colon refer to bonds to the central atom occurring immediately after the bridging group in the CRU or in the polymer chain. Multiple bridging groups between the same pair of central atoms are cited in alphabetical order, each preceded by the Greek letter μ, with all set off by appropriate enclosing marks to reduce the possibility of misinterpretation.

The name of the CRU is then constructed by citing the names of the subunits, starting with the senior subunit and proceeding in the direction of decreasing seniority as determined according to Section 7.2.2. *See* the examples of the complete names for polymers in the remainder of this chapter for examples of how constituent subunits and CRUs are named.

## 7.3 Polymer Nomenclature

### 7.3.1 Polymers with One Central Atom in the CRU

The name for the CRU, which consists of the name for the central atom and any associated side groups followed by the name(s) for the bridging group(s), is inserted into the general polymer name (*see* Section 7.1).

Examples:

$-\!\left(\mathrm{S}\right)\!_n-$ *catena*-poly[sulfur]

$-\!\left(\mathrm{Sn(CH_3)_2}\right)\!_n-$ *catena*-poly[dimethyltin]

$-\!\left(\mathrm{Ag{-}NC}\right)\!_n-$ *catena*-poly[silver-μ-(cyano-*N:C*)]
[*see* Section 7.2.3 (3) for choice of direction]

$-\!\left(\mathrm{Zn(NH_3)(Cl){-}Cl}\right)\!_n-$ *catena*-poly[(amminechlorozinc)-μ-chloro]

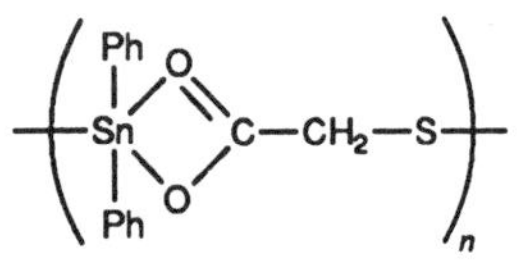

*catena*-poly[(diphenyltin)-μ-[mercapto-acetato(2–)-*O,O′:S*]]
[*see* Section 7.2.3 (1) for choice of direction]

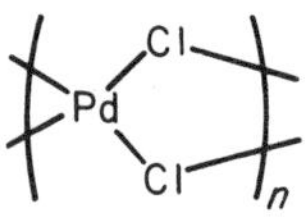

*catena*-poly[palladium-di-μ-chloro]
(Without the prefix "*catena*-", this name does not distinguish between the single-strand formula shown and a two-dimensional polymeric structure.)

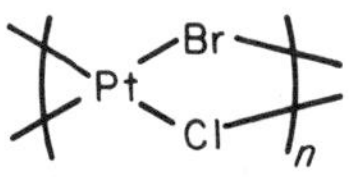

*catena*-poly[platinum-μ-bromo-μ-chloro]

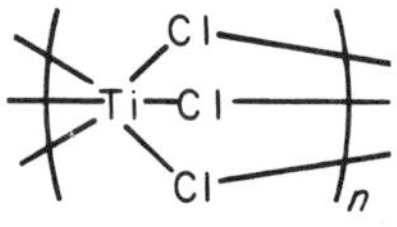

*catena*-poly[titanium-tri-μ-chloro]
(Without the prefix "*catena*-", this name does not distinguish among a number of structural possibilities.)

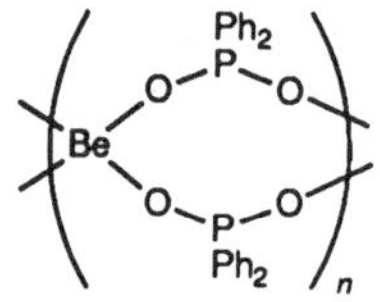

*catena*-poly[beryllium-bis-μ-[diphenyl-phosphinato(1–)-*O:O′*]]

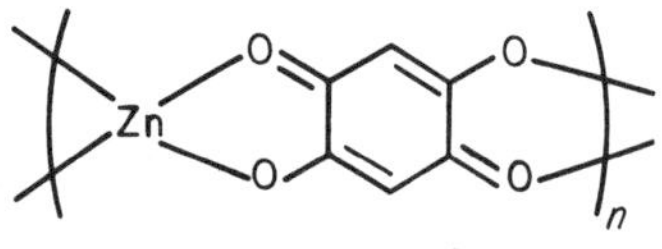

*catena*-poly[zinc-μ-[2,5-dihydroxybenzo-quinonato(2–)-$O^1,O^2:O^4,O^5$]]

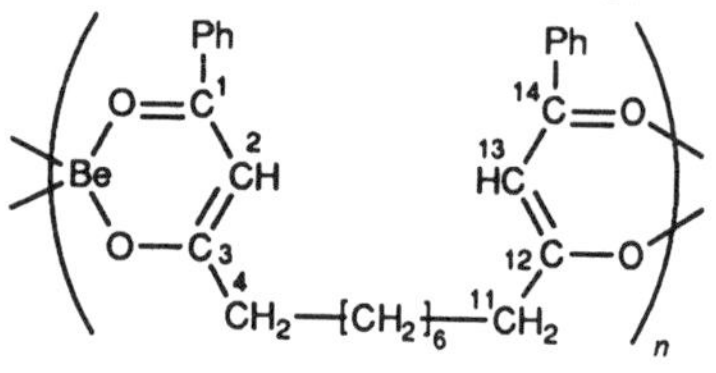

*catena*-poly[beryllium-μ-[1,14-diphenyl-1,3,12,14-tetradecanetetronato(2–)-$O^1,O^3:O^{12},O^{14}$]]

## 7.3.2 Polymers with More Than One Central Atom in the CRU

*7.3.2.1 Central Atoms All the Same.* The senior central atom is chosen according to the appropriate criteria in Section 7.2.2, and the direction of citation is chosen according to Section 7.2.3.

Examples:

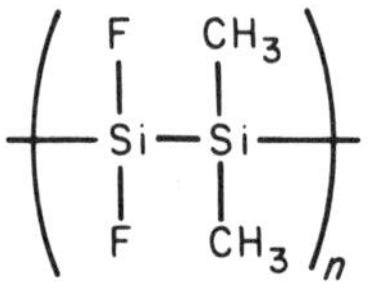

*catena*-poly[(difluorosilicon)(dimethylsilicon)]
[seniority based on Section 7.2.2 (5)]

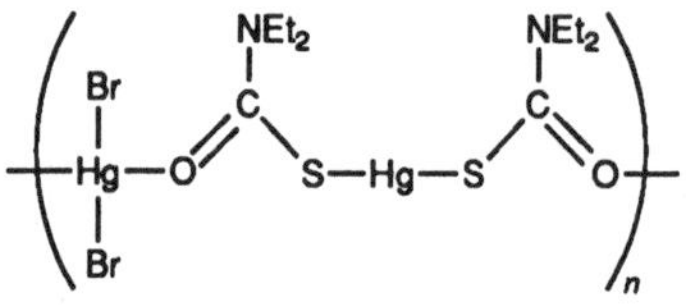

*catena*-poly[(dibromomercury)-μ-(diethylthiocarbamato-*O:S*-mercury-μ-(diethylthiocarbamato-*S:O*]
[seniority based on Section 7.2.2 (4); direction based on Section 7.2.3 (3)]

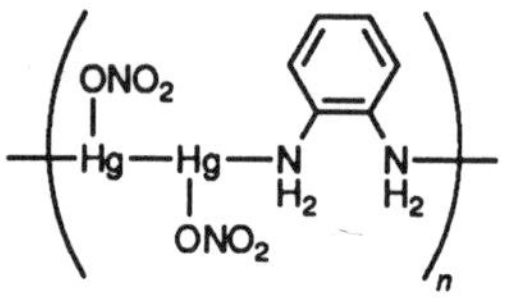

*catena*-poly[(nitratomercury)(nitrato-mercury)-μ-(1,2-benzenediamine-*N:N′*)]
[direction based on Section 7.2.3 (2)]

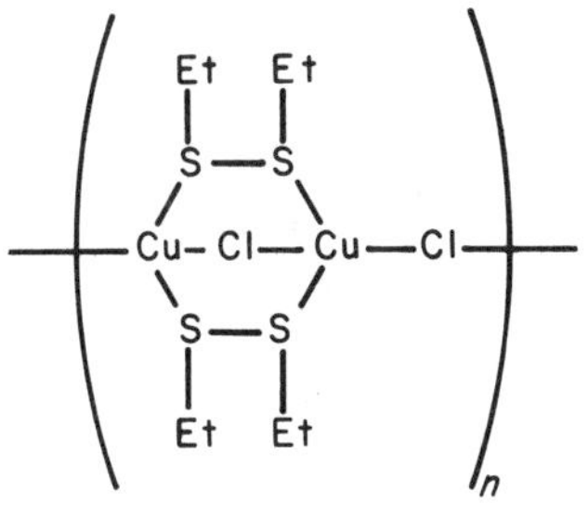

*catena*-poly[copper-[μ-chloro-bis-μ-(diethyl disulfide-*S:S′*)]-copper-μ-chloro]
[direction based on Section 7.2.3 (1)]

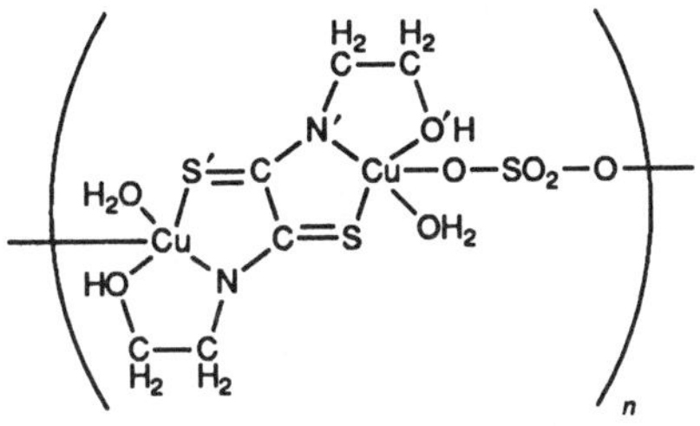

*catena*-poly[(aquacopper)-μ-[*N:N′*-bis(2-hydroxyethyl)dithiooxamido(2–)-*N,O,S′:N′,O′,S*](aquacopper)-μ-[sulfato(2–)-*O:O′*]]
[direction based on Sections 7.2.3 (1) and (3)]

*7.3.2.2 **Different Central Atoms.*** The senior central atom is chosen according to Section 7.2.2 (2), and the direction of citation is chosen according to Section 7.2.3.

Examples:

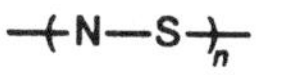

*catena*-poly[nitrogen-μ-thio]

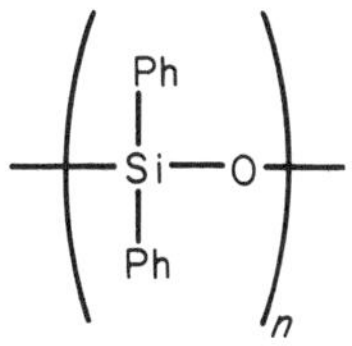

*catena*-poly[(diphenylsilicon)-μ-oxo]

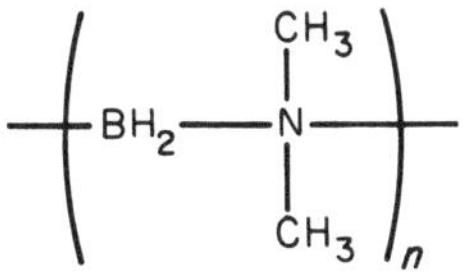

*catena*-poly[(dihydroboron)-μ-(dimethylamido)]

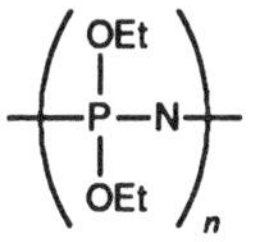

*catena*-poly[(diethoxophosphorus)-μ-nitrido]

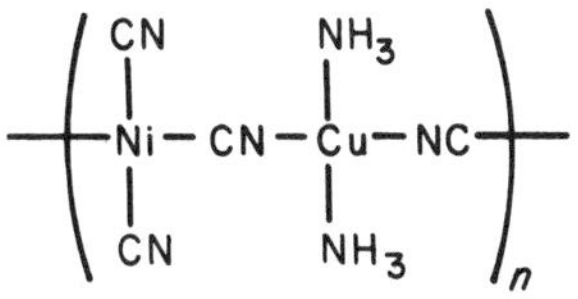

*catena*-poly[[bis(cyano-*C*)nickel]-μ-(cyano-*C:N*)-(diamminecopper)-μ-(cyano-*N:C*)]

*7.3.2.3 **Polynuclear Centers.*** Polymers with one polynuclear center are named in the same manner as polymers with one mononuclear center. The polynuclear center is the senior subunit and, hence, is cited first in naming the CRU. Both positions of the polynuclear center to which bridging groups are attached are indicated by numerical locants that are inserted between the name of the polynuclear center and the name(s) of the bridging group(s). The locants that refer to the first polynuclear center in the CRU are cited first and separated by a colon from the locants for the next polynuclear center in the chain. Numbering rules for polynuclear species have not yet been officially defined, so the numbering shown here is based on the system discussed in Section 6.4.1.

Examples:

$\{[octahedro\text{-}W_6(\mu\text{-}Br)_8(2,3,4,5\text{-}Br_4)](Br_4)\}_n$ — *catena*-poly[(octa-μ-bromo-2,3,4,5-tetrabromo-*octahedro*-hexatungsten)-6:1-μ-[tetrabromido(2–)]]

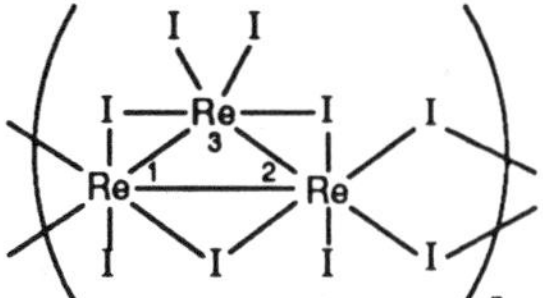


*catena*-poly[[1,2:1,3:2,3-tri-μ-iodo-1,2,3,3-tetraiodo-*triangulo*-trirhenium-(3 *Re–Re*)]-2,2:1,1-di-μ-iodo]

## 7.3.3 Polymers with Ionic Repeating Units

Polymers with ionic CRUs are named in the same general manner as described in the preceding sections. The charge on the CRU may be indicated by the Ewens–Bassett number (*see* Section 2.3.3) cited after the name of the ionic CRU. Stock numbers (*see* Section 2.3.2) may be used to denote the oxidation number of the central atom by attachment to the name of the central atom in the usual manner.

Examples:

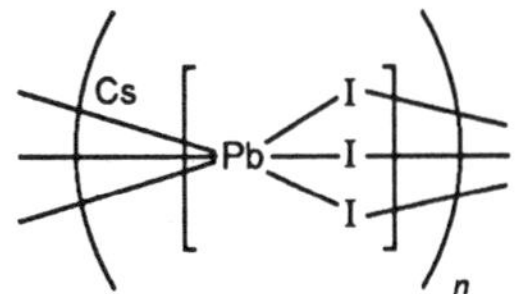


*catena*-poly[cesium [plumbate-tri-μ-iodo](1–)]
*catena*-poly[cesium [plumbate(II)-tri-μ-iodo]]

Comment: It might be more meaningful to consider this example to be a salt with a polymeric anion, that is,

$nCs^+$ $[Pb(\mu\text{-}I)_3]_n^{n-}$

cesium *catena*-poly[[plumbate-tri-μ-iodo](1–)]
cesium *catena*-poly[plumbate(II)-tri-μ-iodo]

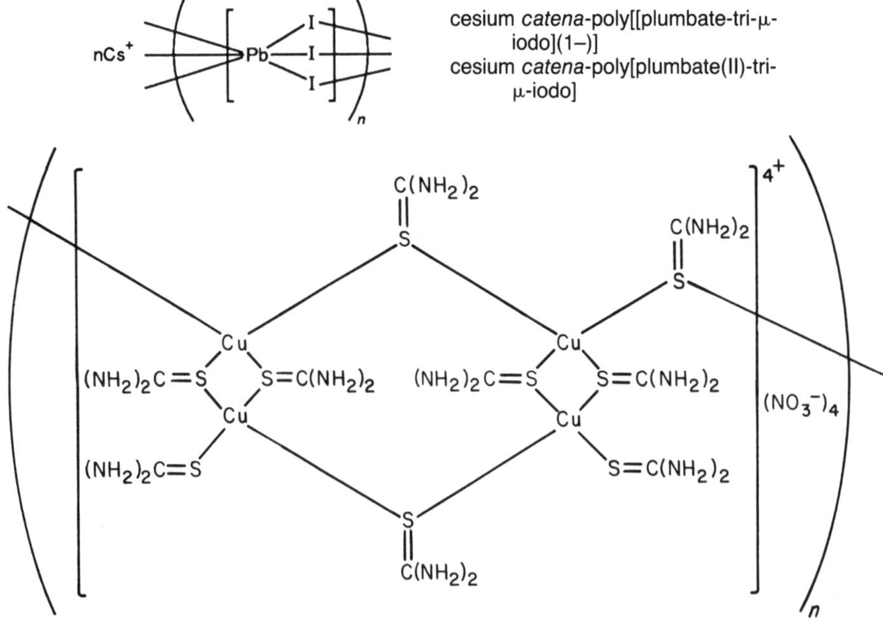


*catena*-poly[{[bis-μ-(thiourea-*S*,*S*)-2-(thiourea-S)di-copper(I)]-1,2:1,2-[bis-μ-(thiourea-*S*:*S*)][bis-μ-(thiourea-*S*,*S*)-2-(thiourea-*S*)dicopper(I)]-1:1-μ-(thiourea-*S*:*S*)} tetranitrate]
*catena*-poly[[bis-μ-(thiourea-*S*,*S*)-2-(thiourea-S)di-copper]-1,2:1,2-[bis-μ-(thiourea-*S*:*S*)]-[bis-μ-(thiourea-*S*,*S*)-2-(thiourea-*S*)di-copper]-1:1-μ-(thiourea-*S*:*S*)(4+)] nitrate

## 7.3.4 End Groups

The end groups of linear inorganic or semiinorganic polymers are indicated by prefixes placed in front of the name of the polymer in a manner analogous to that used for organic polymers (*54*). The end group attached to the first constituent subunit of the preferred CRU (i.e., the senior subunit written as the left terminal subunit in the CRU) is given its ligand name as a side group and designated by the Greek letter α. The end group attached to the other terminal subunit of the preferred CRU is named as a side group if attached to a central atom or, if attached to a bridging group, as a central atom with associated side groups by the usual principles of additive nomenclature and designated by the Greek letter ω.

Example:

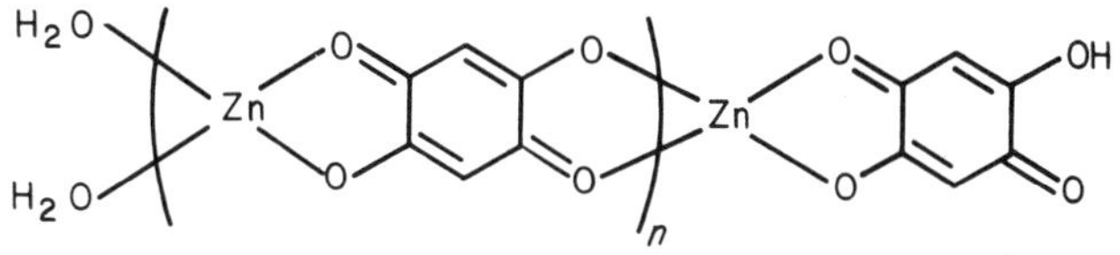

α,α-diaqua-ω-{[2,5-dihydroxy-*p*-benzoquinonato(1–)-*O*[1],*O*[2]]zinc}-*catena*-poly[zinc-μ-[2,5-dihydroxy-*p*-benzoquinonato(2–)-*O*[1],*O*[2]:*O*[4],*O*[5]]]

When either end group may be pictured as being attached to the senior subunit, the end group with the name occurring first in alphabetical order is designated the α end group.

Examples:

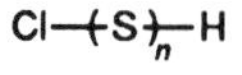

$Cl{-}(S)_n{-}H$ α-chloro-ω-hydrido-*catena*-poly[sulfur]

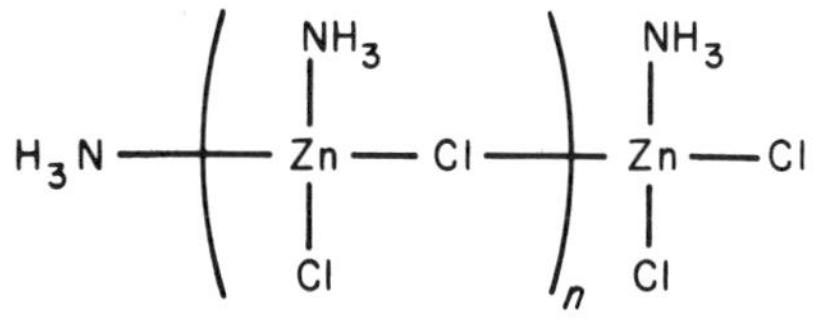

α-ammine-ω-(amminedichlorozinc)-*catena*-poly[(amminechlorozinc)-μ-chloro]
(not α-chloro-ω-(diamminechlorozinc)-*catena*-poly[(amminechlorozinc)-μ-chloro])

End groups that may be considered ionic are named in the usual manner, and the amount of charge is indicated by a Ewens–Bassett number cited at the end of the complete polymer name.

Examples:

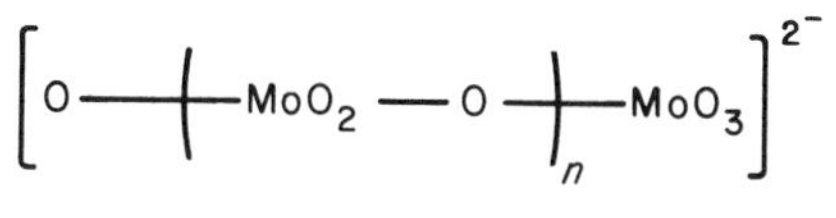

$[O{-}(MoO_2{-}O)_n{-}MoO_3]^{2-}$ {α-oxo-ω-(trioxomolybdenum)-*catena*-poly[(dioxomolybdenum)-μ-oxo]}(2–)

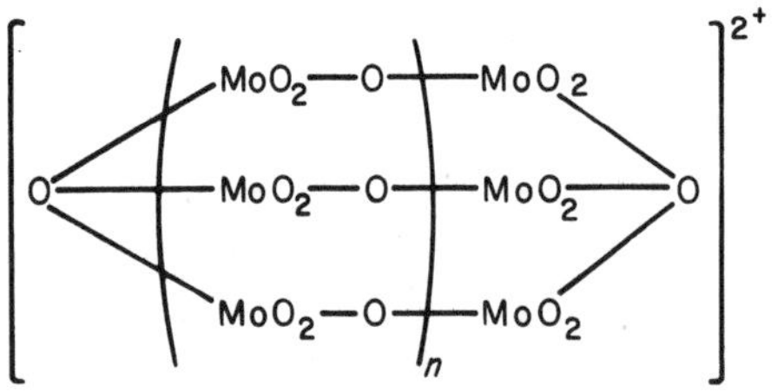

{α-$\mu_3$-oxo-ω-[$\mu_3$-oxo-tris(dioxomolybdenum)]-tris{*catena*-poly[(dioxomolybdenum)-μ-oxo]}}(2+)

The polymer in the second example is a regular single-strand polymer consisting of three chains linked by oxo end groups.

### 7.3.5 Stereochemistry

The stereochemical configuration of a CRU consisting of one central atom with associated groups and one bridging group may be designated by suitable prefixes cited before the appropriate complete polymer name.

Examples:

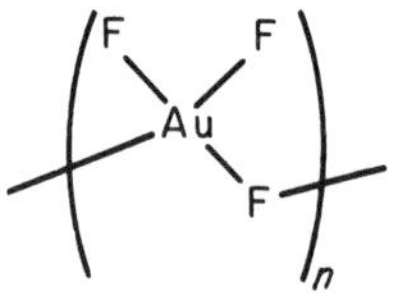

*cis-catena*-poly[(difluorogold)-μ-fluoro]

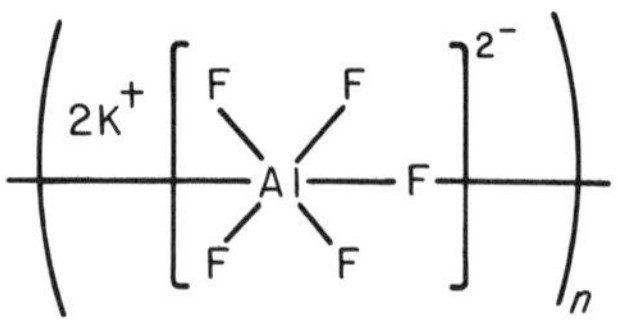

*trans-catena*-poly[dipotassium {[(tetrafluoroaluminate)-μ-fluoro](2–)}]
*trans-catena*-poly[dipotassium {[tetrafluoroaluminate(III)]-μ-fluoro}]

The names in these two examples are constructed in accord with the stereochemical notation recommended for organic polymers. Inorganic nomenclature practice would lead to insertion of the stereochemical prefix directly in front of the name of the CRU, as shown in the following names:

*catena*-poly[*cis*-(difluorogold-μ-fluoro)]
potassium *catena*-poly[*trans*-(tetrafluoroaluminate-μ-fluoro)(2–)]

# Chapter 8
# Acids, Bases, and Their Derivatives

## 8.1 Introduction

Much of early inorganic chemistry was concerned with relatively simple ionic compounds and their constituent ions. These compounds, which were divided into acids, bases, and salts, were given semisystematic names that indicated functionality as well as composition. Their names were for the most part based on the principles originally intended for binary compounds (*see* Section 4.4). These principles were extended to compounds containing more than two elements by treating groups of atoms as single units (*see* pseudobinary names, Section 4.3.1). Although the names generated are not truly systematic, they are still in common use in spite of the fact that there are more systematic procedures for naming all of these substances. Historical methods for naming these classes of compounds are described in this chapter. Additive nomenclature (*see* Section 5.4), however, is preferable, and thus the methods described in this chapter should not be used for naming new compounds. These methods are only recommended for compounds specifically included in the discussion and related tables.

## 8.2 Acids

### 8.2.1 Introduction

In general, functional names are disappearing from inorganic chemistry. Many compounds that are acids according to some definitions do not fit into the classical concept of acids (i.e., compounds that supply $H^+$) and are not named as acids. The following discussion applies only to classical acids.

### 8.2.2 Binary Acids

Classical binary acids contain hydrogen and one other element and are named as binary compounds of hydrogen. The binary hydrides of the halogens have long been called "hydro...ic" acids, where ... represents the stem name for the halogen, in recognition of the acidic nature of their aqueous solutions. Such names, however, should be used only for the aqueous solutions, not for the hydrides themselves.

Examples:

$H_2S$ hydrogen sulfide

HCl hydrogen chloride
("Hydrochloric acid" refers to an aqueous solution of HCl.)

### 8.2.3 Mononuclear Acids Containing Polyatomic Anions

Most of the common classical acids are oxo acids in the inorganic sense; that is, their acidic hydrogens are formally attached to oxygen atoms bound to central atoms. {In organic chemistry, the term "oxo acid" is used to refer to carboxylic acids containing an aldehydic group and/or one or more ketonic groups in the principal chain or parent ring system (*34d*)}.

The names of oxo acids, which fit no one pattern of nomenclature completely, are formed according to the method introduced in 1789 (*8*) in which the suffixes -ous and -ic are used to indicate the degree of saturation (i.e., oxygen content) of the acid. This method is only relative (i.e., -ic meaning higher and -ous lower) and thus does not specify either the number of oxygen atoms or the number of acidic (or nonacidic) hydrogen atoms in either acid. The number of oxygen atoms may vary from one central atom to another for each type of acid, and the oxidation number of the central atom is not necessarily the same within each family of elements. The trivial names and functional structural formulas of the -ous/-ic pairs of mononuclear acids are listed in Table A.V. The -ous/-ic distinction has not been applied to the organic chalcogen oxo acid suffixes. Thus, $-S(O)_2(OH)$ is -sulfonic acid, and -S(O)(OH) is -sulfinic acid, when expressed as a suffix. However, the name -sulfonous acid could be used for -S(O)(OH) without ambiguity, and the names sulfonic and sulfonous acid could be used for the hypothetical acids $(H)S(O)_2(OH)$ and(H)S(O)(OH), respectively. Systematic additive names for the organic chalcogen oxo acids are somewhat longer than the trivial names and are only sparingly used.

The prefixes per- and hypo- attached to the names of the mononuclear -ic and -ous halogen oxo acids, respectively, describe oxo acids with the halogens in a higher and a lower oxidation state, respectively, than in the -ic and -ous acids.

| Examples: | $ClO_3(OH)$ | perchloric acid |
|---|---|---|
| | Br(OH) | hypobromous acid |

The prefix per- has been used with the same meaning in the names of the mononuclear acids of some metallic elements, especially the Group 7 metals. For example, permanganic acid ($HMnO_4$) contains Mn(VII), whereas manganic acid ($H_2MnO_4$) contains Mn(VI). The prefix per- is not to be confused with the prefixes peroxo- and peroxy-, which indicate the presence of a peroxo group (*see* Section 4.4.2). Except for hypophosphorous acid, a traditional name for $H_3PO_2$ whose tautomeric forms are named phosphonous acid and phosphinic acid, the prefix hypo- has not been extended with the same meaning to elements in other families. For example, $S(OH)_2$ is called sulfoxylic acid (*see* paragraph after next) and not hyposulfurous acid.

The prefixes ortho- and meta- have been used to distinguish between two mononuclear oxo acids of a central element in the same oxidation state that differ in the number of water molecules formally required to convert the oxide to the respective acids. The prefix ortho- is not used entirely consistently, referring at times to the "completely hydrated" acid and at times to the acid with two hydroxyl groups more than the meta acid.

| Examples: | $B(OH)_3$, $[B(O)(OH)]_n$ | orthoboric acid[a], metaboric acid |
|---|---|---|
| | $C(OH)_4$, $C(O)(OH)_2$ | orthocarbonic acid, metacarbonic acid[b] |
| | $Si(OH)_4$, $[Si(O)(OH)_2]_n$ | orthosilic acid[a], metasilic acid |

| | |
|---|---|
| $N(O)(OH)_3$, $N(O)_2(OH)$ | orthonitric acid, metanitric acid[b] |
| $P(O)(OH)_3$, $[P(O)_2(OH)]_n$ | orthophosphoric acid[a], metaphosphoric acid |
| $S(OH)_4$, $S(O)(OH)_2$ | orthosulfurous acid, metasulfurous acid[b] |
| $Te(OH)_6$, $Te(O)_2(OH)_2$ | orthotelluric acid[c], metatelluric acid[b] |
| $I(O)(OH)_5$, $I(O)_3(OH)$ | orthoperiodic acid[c], metaperiodic acid[b] |

[a]The prefix ortho- is almost always omitted in practice.

[b]The prefix meta- is almost always omitted in practice.

[c]There is a difference of two molecules of water between the formulas for the ortho and meta oxo acids of tellurium(VI) and iodine(VII), in contrast to the difference of one molecule of water for the other pairs.

Trivial names for other mononuclear oxo acids, such as sulfoxylic acid for $S(OH)_2$, are listed in Table A.V. Some of the species included are hypothetical parent molecules that are used in naming substituted derivatives. Oxo acids with trivial names may also be named by additive nomenclature as anions consisting of central atoms and oxo side groups plus hydrogen cations or as neutral species consisting of central atoms with oxo and hydroxo side groups (*see* Section 5.4). For example, sulfuric acid ($H_2SO_4$) could be named hydrogen tetraoxosulfate(2–), hydrogen tetraoxosulfate(VI), dihydroxodioxosulfur, or dihydroxodioxosulfur(VI).

Mononuclear acids that contain polyatomic anions but that are not classified as oxo acids are named either as pseudobinary compounds of hydrogen if they give rise to -ide anions (*see* Section 4.3.1) or as compounds with hydrogen cations by additive procedures (*see* Section 5.4). Alternative names for the latter formed by combining the suffix -ic with the additive name for the central atom and its associated groups and adding the word "acid" have been used, but such names are not preferred (*22m*).

| Examples: | | |
|---|---|---|
| | HCN | hydrogen cyanide |
| | $HN_3$ | hydrogen azide<br>(preferred to hydrazoic acid) |
| | $H[AuCl_4]$ | hydrogen tetrachloroaurate(1–)<br>hydrogen tetrachloroaurate(III)<br>(preferred to tetrachloroauric(III) acid) |
| | $H_4[Fe(CN)_6]$ | hydrogen hexacyanoferrate(4–)<br>hydrogen hexacyanoferrate(II)<br>(preferred to hexacyanoferric(II) acid) |
| | $H_2[SiF_6]$ | hydrogen hexafluorosilicate(2–)<br>hydrogen hexafluorosilicate(IV)<br>(preferred to hexafluorosilicic acid or fluorosilicic acid) |
| | $H[B(Ph)_4]$ | hydrogen tetraphenylborate(1–)<br>hydrogen tetraphenylborate(III)<br>(preferred to tetraphenylboric acid) |

### 8.2.4 Functional Replacement Analogs of Mononuclear Oxo Acids

Both prefixes and infixes have been used to indicate the replacement of oxygen atoms or hydroxy groups of oxo acids with other atom or groups. The 1957 and 1970 IUPAC Inorganic Nomenclature Rules emphasize the prefix method (*22n*), but they recognize the infix method (*22o*). Section C of the 1979 IUPAC Organic Nomenclature Rules uses both methods, but not as alternatives. Provisional Section D of the IUPAC Organic Nomenclature Rules, which was a joint effort of CNIC and the IUPAC Commission on the Nomenclature of Organic Chemistry (CNOC), included the infix method for mononuclear oxo acids of phosphorus and arsenic (*34e*), but prefix names were provided for many examples. In 1957 CAS began to use the infix method for mononuclear phosphorus oxo acids introduced by the ACS Nomenclature Committee in 1952 (*55*) and extended it to mononuclear arsenic and carbon oxo acids in 1972 (*32b*).

The affixes listed in Table A.VII are joined, as either prefixes or infixes, with the trivial names of the oxo acids. Two or more affixes to the same name are cited in alphabetical order. Prefix names are generally used when oxygen is replaced by another chalcogen or a hydroxyl group by a halogen. [Some names derived in this fashion are identical to the additive "acid" names that are deemed less desirable than the additive "cationic hydrogen" names (*see* Section 8.2.3)]. An infix, which is inserted before the -ic or -ous in the acid name, is preceded by "o" in the names for derivatives of phosphorus, etc., acids. If the replacement product is no longer an acid, it is named according to Chapter 4 or 9, but the names for nitrogen derivatives are derived by replacing "acid" in the name with "amide" or "imide" as appropriate.

Examples:

| | |
|---|---|
| $S(O)(S)(OH)_2$ | thiosulfuric acid[a] |
| $P(F)(O)(OH)_2$ | fluorophosphoric acid[a]<br>phosphorofluoridic acid |
| $H[PF_6]$ | hexafluorophosphoric acid<br>(cf. Section 8.2.3) |
| HSeCN | selenocyanic acid |
| $As(S)(SH)_3$ | tetrathioarsenic acid<br>arsenotetrathioic acid |
| $HBF_2(O) \cdot H_2O$ | difluoroboric acid |
| $S(O)_2(NH_2)(OH)$ | sulfuramidic acid<br>(often shortened to sulfamic acid or to sulfamidic acid)<br>amidosulfuric acid |
| $C(O)(NH_2)(OH)$ | carbonamidic acid<br>(usually shortened to carbamic acid) |
| $C(N_3)(S)(SH)$ | azidodithiocarbonic acid<br>carbonazidodithioic acid |
| PH(OH)(Cl) | chlorophosphonous acid[b]<br>phosphonochloridous acid |
| $P(O)(NH_2)_3$ | phosphoric triamide<br>phosphoryl triamide<br>(often shortened to phosphamide) |
| $S(O)_2(NH)$ | sulfuric imide<br>sulfonyl imide<br>[often shortened to sulfimide, the IUPAC name for $H_2S(NH)$] |
| $(PCl_2N)_3$ | trimeric phosphorus dichloride nitride<br>trimeric phosphonitridic dichloride<br>(often called trimeric phosphonitrile chloride) |

[a]The prefix mono- is occasionally encountered. *See* Table A.II, footnote *a*.

[b]The prefix method can be ambiguous when the same morpheme is used for an atom or group that is replacing either a hydroxyl group or a nonacidic hydrogen. For example, this name could describe $ClP(OH)_2$ as well as PH(OH)Cl (*see* Section 8.2.5).

### 8.2.5 Substituted Mononuclear Acids

Substitution for nonacidic hydrogen atoms in oxo acids and their derivatives is indicated by appropriate substitutive prefixes of organic nomenclature (*34*). Even if the parent acid with nonacidic hydrogen atoms is not known, substitutive names are based on the name of the hypothetical parent acid in some cases (such as some acids of arsenic and phosphorus) (*see* Table A.V). Although there are names for similar hypothetical parent acids based on sulfur, they are generally not used for substitutive names.

Two methods are used to incorporate these prefixes into names. All prefixes are cited in alphabetical order in front of the name of the parent acid when the infix method is used (*34f, 55*), and italicized element symbols are used to locate positions of substituents. (Arabic numbers are often used with the infix "hydrazido".) When a replacement analog of an oxo acid is named by the prefix method, substituents not on the central atom have been included with the appropriate prefix (*34f, 34g*). Although the latter method may avoid the use of some locants, its principles are not fully described and are not consistent with principles for creating alphabetical listings. Additional study is needed.

Examples:

|  |  |
|---|---|
| $(Ph)_2P(O)(OH)$ | diphenylphosphinic acid |
| $CH_3P(O)(OH)(OCN)$ | methylphosphonocyanatidic acid |
| $(Ph)P(NHCH_3)(SH)$ | variants of prefix method:<br>(methylamido)(phenyl)thiophosphonous acid<br>*N*-methyl-*P*-phenyl(amidothiophosphonous acid)<br>infix method:<br>*N*-methyl-*P*-phenylphosphonamidothious acid |
| $S(=NCH_3)_2(Cl)(OH)$ | variants of prefix method:<br>chlorobis(methylimido)sulfuric acid<br>dimethyl(chlorodiimidosulfuric acid)<br>infix method:<br>dimethylsulfurochloridodiimidic acid |

## 8.2.6 Homopolynuclear Oxo Acids

Polynuclear oxo acids that may be considered to be formally derived by condensation of molecules of the same polybasic mononuclear oxo acid with loss of water molecules are named by attaching a type-1 numerical prefix (*see* Table A.II) to the name of the polybasic mononuclear oxo acid. This prefix indicates the number of mononuclear oxo acid molecules involved. The prefix pyro- has been used in the past instead of di- and is still occasionally found. Cyclic and chain structures may be distinguished by the italicized prefixes *cyclo*- and *catena*-.

Examples:

|  |  |
|---|---|
| $(HO)S(O)_2OS(O)_2(OH)$ | disulfuric acid |
| $(HO)_2P(O)OP(O)(OH)_2$ | diphosphoric acid |
| $(HO)P(H)(O)OP(H)(O)(OH)$ | diphosphonic acid |
| $(HO)_3SiOSi(OH)_2OSi(OH)_3$ | trisilicic acid |
| $(HO)Cr(O)_2OCr(O)_2OCr(O)_2OCr(O)_2(OH)$ | tetrachromic acid |
| (structure) | tetraboric acid |
| (structure) | *cyclo*-triphosphoric acid |
| $(HO)_2P(O)OP(O)(OH)OP(O)(OH)_2$ | triphosphoric acid<br>*catena*-triphosphoric acid |

```
HO   O—B
  \B/ O \O
   O |   |
   O  \  B
    \B—O  \OH
```

```
         OH
         |
      O—PO
     /     \
HO(O)P      O
     \     /
      O—PO
         |
         OH
```

The name pyrosulfurous acid, used traditionally for an acid with the composition $H_2S_2O_5$, was changed to disulfurous acid along with the general replacement of the prefix pyro- with di-, even though the structure is known to be $(HO)(O)_2SS(O)(OH)$ rather than $(HO)(O)SOS(O)(OH)$. The name pyrosulfurous acid should be retained for the former structure, and the name disulfurous acid should be used only for the latter. Another name, hypodisulfuric(IV,VI) acid, can be generated for the former structure by using the method for naming binuclear acids in which the central atoms are directly linked (*see* discussion later in this section).

Replacement of oxygen atoms or hydroxy groups of polynuclear oxo acids may be indicated by the prefixes of Table A.VII (*see* Section 8.2.4) (infixes have not been used). Serially primed italicized element symbols or locant numbers are used as needed to denote the central atom at which replacement occurs. The bridging symbol μ with preceding locants, if needed, is used to indicate replacement of bridging oxygen atoms. The order for citation and preference of bridging versus nonbridging replacement has not been defined.

Examples:

|  |  |
|---|---|
| $(H_2N)S(O)_2OS(O)_2(OH)$<br>(locants: S 1, S 2) | *S*-amidodisulfuric acid<br>1-amidodisulfuric acid |
| $(HO)_2P(O)NHP(O)(OH)_2$ | μ-imidodiphosphoric acid |
| $(HO)(Cl)P(O)OP(O)(Cl)(OH)$<br>(locants: P 1, P 2) | *P,P′*-dichlorodiphosphoric acid<br>1,2-dichlorophosphoric acid |
| $(HO)C(O)OOC(O)(OH)$ | μ-peroxodicarbonic acid |
| $(HOO)C(O)OC(O)(OH)$<br>(locants: C 1, C 2) | *C*-peroxodicarbonic acid<br>1-peroxodicarbonic acid |
| $(H_2N)_2P(O)NHP(O)(SH)(OH)$<br>(locants: P 1, P 2) | *P,P*-diamido-μ-imido-*P′*-thiodiphosphoric acid<br>1,1-diamido-μ-imido-2-thiodiphosphoric acid |
| $(HO)_2P(O)OOP(NH)(OH)OP(O)(OH)_2$<br>(locants: P 1, P 2, P 3) | *P′*-imido-μ-peroxotriphosphoric acid<br>2-imido-μ-peroxotriphosphoric acid |
| $(HO)(Ph)P(S)OP(O)(Ph)(OH)$<br>(locants: P 1, P 2) | *P,P′*-diphenyl-*P*-thiodiphosphonic acid<br>1,2-diphenyl-1-thiodiphosphonic acid |

Although there are no fully accepted general rules for naming variants of homopolynuclear acids, certain procedures have been used for species in which the central element is present in different oxidation states and in which the central atoms are directly linked. The former type has been named as "poly...ic acids", with the oxidation numbers of the atoms of the central element given by their Stock numbers appended to the end of the "poly...ic" name in the order of occurrence in the chain and enclosed in parentheses. If there is a choice, lower oxidation numbers are preferred for earlier citation and numbering. This type of homopolynuclear acid can also be named by the method used for mixed anhydrides (*see* Section 8.5.1.4).

Examples:

|  |  |
|---|---|
| $(HO)_2POP(O)(OH)_2$ | diphosphoric(III,V) acid |
| $(HO)_2POP(O)(OH)OP(OH)OP(O)(OH)_2$ | tetraphosphoric(III,V,III,V) acid |
| $(HO)_2POP(O)(OH)(NH_2)$<br>(locants: P 1, P 2) | *P′*-amidodiphosphoric(III,V) acid<br>2-amidodiphosphoric(III,V) acid |

The prefix hypo- attached to the name of a mononuclear oxo acid has been used to describe a symmetrical binuclear oxo acid in which the central atoms are directly linked. This use is presumably justified by the practice of ignoring direct bonds between like atoms in the determination of oxidation number. It has, however, a different meaning than the hypo- in hypochlorous acid (HClO) and hypophosphorous acid ($H_3PO_2$) (*see* Section 8.2.3), so we recommend that the prefix, when used in this sense, be hypodi- rather than simply hypo- to avoid confusion.

Examples:

|  |  |
|---|---|
| $(HO)_2P(O)P(O)(OH)_2$ | hypodiphosphoric acid[a]<br>hypophosphoric acid[b] |
| $(HO)_2PP(OH)_2$ | hypodiphosphorous acid[c] |
| $(HO)(H)P(O)P(O)(H)(OH)$ | hypodiphosphonic acid[a]<br>hypophosphonic acid[b] |

| | |
|---|---|
| (HO)(H)AsAs(H)(OH) | hypodiarsonous acid[a]<br>hypoarsonous acid[b] |

[a]Name we recommend.

[b]Traditional name (method of this section).

[c]The name hypophosphorous acid cannot be used for this species because it has been given to the mononuclear acid $H_3PO_2$.

Some polynuclear oxo acids of sulfur consist of a chain of bivalent sulfur atoms connecting two sulfonic or two sulfinic acid groups. These acids may be named by attaching a type-1 numerical prefix (*see* Table A.II) that expresses the total number of sulfur atoms to the name thionic or thionous acid, according to whether the terminal acid groups are sulfonic or sulfinic acid groups, respectively. When such compounds contain three or more sulfur atoms, substitutive nomenclature based on an appropriate parent hydride (*see* Section 9.2) is an alternate method for naming them.

Examples:

| | |
|---|---|
| $(HO)S(O)_2S(O)_2(OH)$ | dithionic acid[a] |
| (HO)S(O)SS(O)(OH) | trithionous acid<br>sulfanedisulfinic acid |

[a]Dithionic acid could also be named hypodisulfuric acid (*see* preceding paragraph), but "hyposulfuric acid" cannot be used because the term hyposulfate has been used for thiosulfate in photography.

Structural names for more complex homopolynuclear oxo acids may be derived by using the recently published recommendations for naming rings and chains (*56*) and polyanions (*29*).

### 8.2.7 Heteropolynuclear Oxo Acids

***8.2.7.1 General Discussion.*** The nomenclature of heteropolynuclear oxo acids and their oxo anions is one of the areas of nomenclature under active development by CNIC. The drafts under consideration indicate that the procedures in the 1970 IUPAC Inorganic Rules (*22p*) will be almost completely replaced. The basic approach for anions in the 1970 rules is to treat the component binary oxo groups as oxo anions (*see* Section 8.4), select one as the central group, cite it last in the name, and cite the others as anionic side groups. The prefixes *catena*- and *cyclo*- are used as appropriate. The corresponding acids are named as hydrogen salts (*see* Section 8.2.3). Names based on these principles will probably become obsolete soon.

***8.2.7.2 Chains.*** A heterodinuclear oxo acid species is named by first citing the ligand name for the binary oxo group that comes first alphabetically and then citing the anion name for the remaining oxo group.

Examples:

| | |
|---|---|
| $H_4[O_3AsOPO_3]$ | hydrogen arsenatophosphate(4–)<br>(arsenate before phosphate) |
| $H_2[O_3SOCrO_3]$ | hydrogen chromatosulfate(2–)<br>(chromate before sulfate) |

If one of the constituent binary oxo groups in a longer chain is an obvious central oxo group, the other oxo groups are cited as ligands in alphabetical order in front of the anion name for the central oxo group.

Examples:

| | |
|---|---|
| $H_5[O_3AsOP(O)_2OAsO_3]$ | hydrogen bis(arsenato)phosphate(5−) |
| $H_4[O_3CrOP(O)(OSO_3)OAsO_3]$ | hydrogen (arsenato)(chromato)(sulfato)phosphate(4−) |

If there is not an obvious central binary oxo group in a longer chain, the name for the terminal oxo group that comes first alphabetically is cited, followed sequentially by the names for the interior oxo groups. The name ends with the anionic name for the other terminal oxo group, which is arbitrarily considered to be the central group.

Examples:

| | |
|---|---|
| $H_4[O_3CrOP(O)_2OAsO_3]$ | hydrogen (arsenatophosphato)chromate(4−) (arsenate before chromate) |
| $H_4[O_3SOCr(O)_2OAs(O)_2OPO_3]$ | hydrogen [(phosphatoarsenato)chromato]sulfate(4−) (phosphate before sulfate) |
| $H_4[O_3POCr(O)_2OS(O)_2OAsO_3]$ | hydrogen [(arsenatosulfato)chromato]phosphate(4−) (arsenate before phosphate) |

*8.2.7.3* ***Rings.*** The binary oxo group to be cited first in naming a monocyclic ring is the oxo group in the ring that comes first alphabetically by name. The remaining oxo groups are cited in order of occurrence, starting with the adjacent oxo group that is alphabetically earlier. The oxo group to be named as an anion (i.e., the oxo group treated as the central group) is the oxo group adjacent to the starting oxo group in the less preferred direction.

Examples:

hydrogen *cyclo*-arsenatochromatosulfato-phosphate(2−)
(start with arsenato and go toward chromato, not toward phosphato)

hydrogen *cyclo*-arsenatochromatophosphato-sulfatochromate(2−)
(start with arsenato and go through chromato toward phosphato, not sulfato)

*8.2.7.4* ***Three-Dimensional Structures.*** The species for which nomenclature procedures are suggested (*22q*) all consist of a central atom surrounded by condensed octahedrons of $MO_6$ in which M represents Mo, W, or V. The name is constructed by citing the stem of M with the suffix "o" and the appropriate type-1 numerical prefix from Table A.II as a side group of the central atom. If the oxidation number of any atom has to be given, it may be necessary to place it immediately after the name of that atom to avoid ambiguity.

Examples:

| | |
|---|---|
| $H_3[PW_{12}O_{40}]$ | hydrogen dodecatungstophosphate(3−) |
| $H_5[IW_6O_{24}]$ | hydrogen hexatungstoperiodate(5−) |
| $H_6[Mn^{IV}Mo_9O_{32}]$ | hydrogen nonamolybdomanganate(IV)(6−) |
| $H_7[Co^{II}Co^{III}W_{12}O_{42}]$ | hydrogen dodecatungstocobalt(II)cobalt(III)ate(7−) |
| $H_3[PMo_{10}V_2O_{39}]$ | hydrogen decamolybdodivanadophosphate(3−) |

Before the preceding names were developed, compounds of this type were named phosphotungstic acid, etc., in a procedure similar to the method discussed in Section 8.2.7.2. Structural names for more complex heteropolynuclear oxo acids may be derived by using the recently published recommendations for naming rings and chains (*56*) and polyanions (*29*).

## 8.3 Bases

The definition of a base has been so modified that there is no functional group common to a majority of bases. Consequently, there is no general method for naming bases. The hydroxides and oxides of electropositive elements, which constitute a significant fraction of bases, are for the most part named by the methods of Chapter 4 because they are primarily binary or pseudobinary compounds.

Examples:

| | |
|---|---|
| KOH | potassium hydroxide |
| CaO | calcium oxide |

## 8.4 Salts and Esters

### 8.4.1 Neutral Salts

Those compounds known as salts represent a wide variety of types and individual species. The majority of these compounds are given names of two or more words in which the name for the cation (or the names for the cations in alphabetical order if there is more than one) is followed by the name of the anion or (should there be more than one) by the names of the anions in a separate alphabetical order. The only exceptions to alphabetical order are that hydrogen is always cited last among cations (*see* Section 8.4.2) and that hydration of cations, when given, does not disturb the alphabetical order of the cations. The name of a cation is that of the element or the polyatomic entity (*see* Sections 4.3 and 4.4), whereas that for an anion is derived from the name for the corresponding acid by changing the characteristic affixes hydro...ic, -(i)ous, and -ic to -ide, -ite, and -ate, respectively, and eliminating the word "acid" (*see also* Sections 4.3 and 4.4). The stoichiometry is given, when appropriate, by numerical prefixes (*see* Table A.II). Type-1 prefixes are normally used, but it is necessary to use type-2 prefixes to indicate two or more oxo anions in order to avoid confusion with the names of polynuclear oxo anions.

Examples of names for simple and mixed salts follow. Although the formulas for double, triple, etc., salts are often written as though they are addition compounds, the preferred names are based on the constituent ions because such salts are considered to be mixed salts for nomenclature purposes.

Examples:

| | |
|---|---|
| $NiBr_2$ | nickel(2+) bromide<br>nickel(II) bromide |
| $KNa_2PO_4$ | potassium disodium phosphate |
| $CrK(SO_4)_2$ | chromium(3+) potassium bis(sulfate)<br>chromium(III) potassium bis(sulfate) |
| $NaCl \cdot NaF \cdot 2Na_2SO_4$<br>$Na_6ClF(SO_4)_2$ | hexasodium chloride fluoride bis(sulfate) |
| $KCl \cdot MgSO_4$<br>$KMgClSO_4$ | magnesium potassium chloride sulfate |
| $MgNa(UO_2)_3(OAc)_9$ | magnesium sodium triuranyl(2+) nonaacetate |

### 8.4.2 Acid and Basic Salts

Salts that contain acidic hydrogen in addition to other cations or contain oxide or hydroxide anions in addition to other anions have been termed acid salts and basic salts, respectively. They are all preferably named as mixed salts, with hydrogen cited as the last cation and with hydroxide and oxide taking their proper places in the alphabetical sequence of anions. Although the 1970 IUPAC Inorganic Rules state that the word hydrogen is to be attached directly to the name of the anion (*22r*), in the 1979 IUPAC Organic Rules (*34h*) and in the comments to the 1957 IUPAC Inorganic Rules (*21*) hydrogen is written as a separate word because "hydrogen" is a word, not a prefix. (The suggestion that the prefix hydro- be used instead of hydrogen has not found favor.) The names of compounds that contain anions in which there is nonreplaceable hydrogen seldom produce ambiguity about which hydrogens are replaceable. Basic salts have been named as hydroxy or oxy salts (e.g., aluminum oxychloride for AlClO), but such names are not recommended.

Examples:

| | |
|---|---|
| $LiHCO_3$ | lithium hydrogen carbonate |
| $CaHPO_4$ | calcium hydrogen phosphate |
| $CaH_4(PO_4)_2$ | calcium tetrahydrogen bis(phosphate)<br>calcium bis(dihydrogen phosphate) |
| $KH(PHO_3)$ | potassium hydrogen phosphonate |
| MgCl(OH) | magnesium chloride hydroxide |
| AlClO | aluminum chloride oxide |
| $Nb_2O(SO_4)_4$ | diniobium(5+) oxide tetrakis(sulfate)<br>diniobium(V) oxide tetrakis(sulfate) |
| $Cu_4(OH)_6SO_4$ | tetracopper(2+) hexahydroxide sulfate<br>tetracopper(II) hexahydroxide sulfate |

### 8.4.3 Esters

Neutral esters of inorganic acids are given names in the same fashion as salts, with the name of the organic group in place of the name of the cation. In mixed esters the organic groups are cited in alphabetical order. Partial esters of polybasic acids are treated in a manner analogous to acid salts. Salts of partial esters are named by citing the cations before the organic groups. If acidic hydrogens remain, they are cited after the organic groups.

Examples:

| | |
|---|---|
| EtONO | ethyl nitrite |
| $(MeO)_4Si$ | tetramethyl orthosilicate |
| MeP(O)(OMe)(OPh) | methyl phenyl methylphosphonate |
| $MeOS(O)_2(OH)$ | methyl hydrogen sulfate |
| $(EtO)_2P(O)(OH)$ | diethyl hydrogen phosphate |
| $Na[EtOS(O)_3]$ | sodium ethyl sulfate |
| $K[MeOP(O)_2(OH)]$ | potassium methyl hydrogen phosphate |

## 8.5 Other Acid Derivatives

### 8.5.1 Acid Anhydrides

***8.5.1.1 Acids Containing Only Oxo and Hydroxo Groups.*** A fully dehydrated mononuclear inorganic oxo acid [i.e., a compound having no hydrogen atoms attached to oxygen atom(s)] is usually named as a binary oxide, not as an anhydride, and corresponding chalcogen analogs are named as binary chalcogenides, not as chalcoanhydrides. Thus, $SO_3$ is named sulfur trioxide, not sulfuric anhydride, and $N_2S_5$ is named dinitrogen pentasulfide, not dithionitric thioanhydride.

*8.5.1.2 **Monobasic Oxo Acids.*** Symmetrical anhydrides of monobasic oxo acids (or their chalcogen analogs) with at least one atom or side group other than oxo or hydroxo (including hydrogen) attached to the central atom are named by replacing the word "acid" in the name of the acid by the class name "anhydride" (or "chalcoanhydride"). An anhydride has an oxygen atom between central atoms (i.e., ...MOM...), whereas a chalcoanhydride has a chalcogen atom between central atoms (e.g., ...MSM...).

Examples:

| | |
|---|---|
| $H_2P(O)OP(O)H_2$ | phosphinic anhydride |
| ClS(O)SS(O)Cl | chlorosulfurous thioanhydride |
| $PhP(S)(NH_2)OP(S)(Ph)(NH_2)$ | phenylphosphonamidothioic anhydride<br>*P,P'*-diphenyl(amidothiophosphonic anhydride) |
| $Cl_2POPCl_2$ | phosphorodichloridous anhydride<br>dichlorophosphorous anhydride |

*8.5.1.3 **Polybasic Oxo Acids.*** Symmetrical anhydrides of polybasic oxo acids or their chalcogen analogs that still contain acidic hydrogen atoms may be named either as polynuclear oxo (chalco) acid analogs (*see* Section 8.2.6) or by replacing the word "acid" in the name of the parent acid by an appropriate descriptive phrase such as bimolecular anhydride, trimolecular cyclic anhydride, or tetramolecular trianhydride. The terms bimolecular, trimolecular, etc., are usually abbreviated bimol., trimol., etc. (Such names may not be completely unambiguous or structurally specific.) Names based on chains or rings of repeating units (*see* Chapter 10) may also be used.

Examples:

| | |
|---|---|
| $HOS(O)_2OS(O)_2OH$ | disulfuric acid<br>sulfuric bimol. monoanhydride |
| 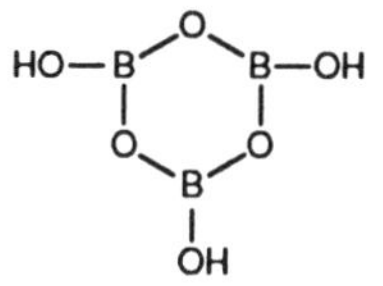 cyclic $(HOBO)_3$ ring: HO—B, O, B—OH; O, O; B—OH | *cyclo*-triboric acid<br>boric trimol. cyclic anhydride |

*8.5.1.4 **Mixed Anhydrides.*** Anhydrides derived from different mononuclear inorganic oxo acids or from inorganic oxo acids and organic acids may be named as polynuclear oxo acids where appropriate (*see* Section 8.2.7) or by citing, in alphabetical order, the first parts of the names of the acid components, followed by a class name such as anhydride or thioanhydride. Appropriate numerical prefixes are used to indicate the number of acid residues involved if there is more than one (i.e., the prefix mono- is not used) and to indicate the number of anhydride linkages (in which case the prefix mono- is used). The second method loses structural specificity quite rapidly as the number and type of acid residues involved increases. It is most useful only for monoanhydrides or polyanhydrides involving one central polybasic acid and two or more other acids.

Examples:

| | |
|---|---|
| $CH_3C(O)OP(O)(OH)_2$ | acetic phosphoric monoanhydride |
| $(HO)_2P(O)OP(OH)_2$ | diphosphoric(III,V) acid<br>phosphoric phosphorous monoanhydride<br>phosphorus(III) phosphorus(V) monoanhydride |
| $CH_3P(O)(OH)OP(O)(OH)_2$ | methylphosphonic phosphoric monoanhydride |

| | |
|---|---|
| $[CH_3C(O)O]_2BOB[OC(O)CH_3]_2$ | tetraacetic diboric tetraanhydride |
| $CH_3P(O)(OH)(OCN)$ | cyanic methylphosphonic monanhydride (*see* example in Section 8.2.5 for name as an acid) |

## 8.5.2 Acid Halides

Acid halides are here defined as derivatives of oxo acids in which all hydroxo groups have been replaced by nonacidic atoms or groups, at least one of which is a halogen atom. They may be named by replacing the word "acid" of the name for the parent acid by the -ide name for the appropriate halogen, or by changing the "...ic acid" ending of the acid name to -oyl or -yl and adding the appropriate halide name as a separate word. Numerical prefixes may be used to denote more than one of the same halogen atom. Different halogen atoms may be cited in alphabetical order, or one may be chosen to be cited as the appropriate class name and the other(s) incorporated into the name of the acid residue by using infixes or prefixes (*see* Section 8.2.5). (Additive names may also be used.)

Examples:

| | |
|---|---|
| $ClS(O)_2Cl$ | sulfuric dichloride<br>sulfuryl dichloride<br>sulfonyl dichloride |
| $C_6H_5P(S)[N(CH_3)_2]Cl$ | (dimethylamido)phenylthiophosphonic chloride<br>*N*,*N*-dimethyl-*P*-phenylphosphonamidothioic chloride<br>(dimethylamido)phenylthiophosphonoyl chloride<br>*N*,*N*-dimethyl-*P*-phenylphosphonamidothioyl chloride |
| $ClS(O)_2NHS(O)_2Br$ | μ-imidodisulfuric bromide chloride<br>μ-imidodisulfuryl bromide chloride |
| $Cl_2POPCl_2$ | diphosphorous tetrachloride<br>(*see* example in Section 8.5.1.2 for names as an anhydride) |

## 8.5.3 Acid Amides and Hydrazides

Derivatives of oxo acids in which all hydroxo groups have been replaced by amido or hydrazido groups may be named in the same way as acid halides (*see* Section 8.5.2) by using the class name amide or hydrazide as appropriate. (Additive names may also be used.)

Examples:

| | |
|---|---|
| $H_2NS(O)_2NH_2$ | sulfuric diamide<br>sulfuryl diamide<br>sulfonyl diamide |
| $P(NH_2)_3$ | phosphorous triamide |
| $C_6H_5P(S)[N(CH_3)_2]_2$ | phenyl(thiophosphonic) bis(dimethylamide)<br>*N*,*N*,*N*′,*N*′-tetramethyl-*P*-phenylphosphonothioic diamide<br>phenyl(thiophosphonoyl) bis(dimethylamide)<br>*N*,*N*,*N*′,*N*′-tetramethyl-*P*-phenylphosphonothioyl diamide |
| $H_2NNHS(O)_2NHNH_2$ | sulfuric dihydrazide<br>sulfuryl dihydrazide<br>sulfonyl dihydrazide |
| $(C_6H_5)_2PNHN(CH_3)_2$ | diphenylphosphinous 2,2-dimethylhydrazide<br>$N^2$,$N^2$-dimethyl-*P*,*P*-diphenylphosphinous hydrazide<br>2,2-dimethyl-*P*,*P*-diphenylphosphinous hydrazide |

Chapter 9

# Substitutive Nomenclature for Covalent Inorganic Compounds

## 9.1 Introduction

Although the primary system for naming inorganic compounds is based on additive nomenclature, substitutive nomenclature is also used. Substitutive procedures, which are based on the methods of organic nomenclature for hydrocarbons, are an alternative method for naming covalent hydrides, particularly those of the nonmetals and semimetals. They are generally confined to the hydrides of boron and the elements of Groups 14, 15, and 16. Substitutive names are used primarily for naming compounds that may be looked upon as derivatives of known or postulated hydrides, not for naming the hydrides themselves. Thus, the principles of substitutive nomenclature developed for hydrocarbons can be extended to a wide variety of noncarbon compounds if such compounds are considered to be derived from molecular hydrides by substitution for hydrogen. Substitutive names are generally shorter than additive names, and their similarity to organic names makes them relatively easy to comprehend.

## 9.2 Mononuclear Binary Hydrides and Their Derivatives

Binary compounds consisting of one central atom and associated hydrogen atoms may be named as salts (*see* Section 4.5.3) or as covalent compounds (*see* Section 4.5.1). Alternatively, mononuclear covalent hydrides of the nonmetals and semimetals may be named as parent hydrides in a method developed by analogy to the methods used for acyclic organic hydrocarbons. In this procedure the names for hydrides with the number of hydrogen atoms equal to the standard bonding number of the element (the number of sigma bonds needed to complete the outer electronic shell of an element) are formed by adding the suffix -ane to a stem name for the element to give the names shown in Table A.VIII.

Hydrides in which the number of hydrogen atoms bonded to the central atom is not equal to the standard bonding number are named by adding the prefix $\lambda^n$- (57) to the appropriate name in Table A.VIII ($n$ is the number of hydrogen atoms bonded to the central atom). Replacement of the hydrogen atoms with other atoms or groups is indicated by appropriate organic substitutive prefixes (*34i*) cited in alphabetical order with suitable numerical prefixes (*see* Table A.II). The organic substitutive name for a given group is not always the same as the inorganic additive name for the same group. These differences are shown in Table A.V.

The trivial names for a few simple binary hydrides [namely, methane ($CH_4$), ammonia ($NH_3$), phosphine ($PH_3$), phosphorane ($PH_5$), arsine ($AsH_3$), arsorane ($AsH_5$), stibine ($SbH_3$), bismuthine ($BiH_3$), and water] are considered acceptable in view of their established use. These names, with the exception of ammonia and water, are suitable for use as the basis for names derived by substitutive nomenclature principles.

Examples:

| | |
|---|---|
| NaH | sodium hydride |
| $LiAlH_4$ | lithium aluminum tetrahydride<br>lithium tetrahydridoaluminate(1−) |
| $SnH_4$ | stannane<br>(*see* Table A.VIII)<br>tin tetrahydride |
| $PHCl_2$ | dichlorophosphane<br>phosphorus dichloride hydride |
| $SiH_2Cl_2$ | dichlorosilane<br>silicon dichloride dihydride |
| $SH_4$ | $\lambda^4$-sulfane |
| $SH_6$ | $\lambda^6$-sulfane |
| $SiH_2$ | $\lambda^2$-silane<br>(*see* Section 9.3 for name as side group) |
| $SiH_3$ | $\lambda^3$-silane<br>(*see* Section 9.3 for name as side group) |
| $SOF_4$ | tetrafluorooxo-$\lambda^6$-sulfane |
| $GeCl_2$ | dichloro-$\lambda^2$-germane |
| $P(S)Cl_3$ | trichlorothioxo-$\lambda^5$-phosphane<br>trichlorothioxophosphorane |

## 9.3 Mononuclear Binary Hydro Groups

Groups that may be derived formally by the removal of one or more hydrogen atoms from a neutral mononuclear molecular hydride may be named by adding a subtractive suffix to the molecular hydride name (*see* Section 9.2) and eliding the final -e of the hydride name, if present, before suffixes starting with -y. Often the entire -ane suffix of the hydride name is elided in forming the names for groups derived from mononuclear molecular hydrides of elements in Group 14 of the periodic table other than carbon, in analogy to the naming of the corresponding hydrocarbon groups. It is likely that CNOC will soon officially redefine some of the subtractive suffixes to be used in naming these groups. The suffixes are expected to be defined in the following fashion:

| | |
|---|---|
| -yl | one hydrogen removed (a monovalent group) |
| -ylidene[a] | two hydrogens removed from the same atom when the two free valences are attached to another single atom (a bivalent group) |
| -diyl[a] | two hydrogens removed from different atoms (a bivalent group); also used instead of -ylidene when the two free valences from a single atom are attached to different atoms |
| -ylidyne | three hydrogens removed from the same atom when the three free valences are attached to another single atom (a trivalent group) |

| | |
|---|---|
| -triyl | three hydrogens removed from different atoms (a trivalent group); also used instead of -ylidyne when the three valences from a single atom are attached to different atoms |

[a]The suffix -ylene has been used for certain bivalent groups derived from hydrocarbons and some mononuclear hydrides (e.g., ethylene for $-CH_2CH_2-$, methylene for $-CH_2-$ or $=CH_2$, and silylene for $-SiH_2-$ or $=SiH_2$). Names such as carbene and nitrene have been used for unattached bivalent groups derived from mononuclear hydrides. The $\lambda^n$ convention (*see* Section 9.2) can also be used to name such unattached biradicals, that is, $\lambda^2$-silane for $SiH_2$ and $\lambda^1$-azane for NH.

Examples:

| | | | | | |
|---|---|---|---|---|---|
| $-CH_3$ | methyl | $-NH_2$ | azanyl | $-SH$ | sulfanyl |
| $-SiH_3$ | silyl | $=SiH_2$ | silylidene<br>silanediyl | $\equiv GeH$ | germylidyne<br>germanetriyl |
| $=PH$ | phosphanylidene<br>phosphinediyl<br>(not phosphinylidene) | $=PH_3$ | phosphoranylidene<br>$\lambda^5$-phosphanylidene | | |

Once again, there are commonly used names:

| | | | | | |
|---|---|---|---|---|---|
| $-NH_2$ | amino | $-PH_2$ | phosphino | $-AsH_2$ | arsino |
| $=NH$ | imino | $=PH$ | phosphinidene | $=AsH$ | arsinidene |
| $-SbH_2$ | stibino | $-BiH_2$ | bismuthino | $-SH$ | mercapto |
| $=SbH$ | stibylene | $=BiH$ | bismuthylene | $-SeH$ | selenyl |

## 9.4 Ions from Mononuclear Molecular Hydrides

Ions may be formed by the formal addition or removal of hydrogen cations or hydride anions to or from molecular hydrides. Names for binary hydro cations formed by the addition of a hydrogen(1+) cation to mononuclear parent hydrides of the Group 15, 16, and 17 elements have customarily been formed by adding the suffix -onium to a stem name for the element (e.g., phosphonium ion for $PH_4^+$). The parallel names "ammonium ion" and "hydronium ion" have been used for $NH_4^+$ and $H_3O^+$, respectively, but the latter is no longer considered acceptable (*30*). Binary hydro cations with the same elemental composition as binary groups may be named by adding the separate word "cation" after the name for the group with appropriate designation of charge (*see* Section 4.5.2) (e.g., methyl cation for $CH_3^+$).

A more general method for naming cations formally derived from mononuclear hydrides by addition of $H^+$ or removal of $H^-$ is to add the suffix -ium or -ylium, respectively, to the -ane name of the molecular hydride, eliding -e before -i or -y. The entire -ane suffix is often elided from the names of Group-14-element hydrides in the same way that it is in the naming of mononuclear binary hydro groups (*see* Section 9.3). The addition of more than one hydrogen cation or the removal of more than one hydride anion may be indicated by a numerical prefix (*see* Table A.II) and suitable locants before the suffix. The use of a Ewens–Bassett number to indicate the charge on the cation [e.g., -ium($n+$)] is not consistent with substitutive nomenclature principles, but names containing such symbolism may occasionally be encountered. Halogen-substituted hydro cations are named according to the principles of substitution.

Examples:

| | |
|---|---|
| $NH_4^+$ | azanium ion |
| $PH_4^+$ | phosphanium ion |
| $CH_3^+$ | methylium ion |
| $BH_2^+$ | boranylium ion |
| $SiH_3^+$ | silylium ion<br>λ²-silanium ion |
| $NF_4^+$ | tetrafluoroazanium ion |
| $Cl_2I^+$ | dichloroiodanium ion |

Binary hydro anions formally derived from neutral mononuclear hydrides by the addition of one or more hydride anions are generally named as coordination entities (*see* Section 5.3). Binary hydro anions formally derived from a neutral mononuclear hydride by the loss of one or more hydrogen cations may be named by adding suffixes such as -ide, -diide, etc., to the -ane name of the molecular hydride, eliding -e before -i. The alternative -ide($n$–), as with the analogous -ium($n$+), is not consistent with substitutive nomenclature principles, but it may be encountered.

Derivatives of hydro anions with names based on parent hydrides may be named according to the principles of substitutive nomenclature. An alternative is to add the prefixes hydro-, dihydro-, or other appropriate prefixes to the name of a monoatomic or homopolyatomic anion with elision of o- before a following vowel. The charge on the anion may be indicated by a Ewens–Bassett number, and the word "anion" or "ion" may be added as a separate term. Binary hydro anions with the same composition as binary groups may be named by adding the separate word "anion" after the name of the binary group and indicating the charge (*see* Section 5.4 for details). A few hydro anions have long-standing trivial names that presumably will not quickly disappear.

Examples:

| | |
|---|---|
| $NH_2^-$ | azanide ion<br>dihydronitride ion<br>azanyl anion<br>amide ion (trivial name) |
| $NH^{2-}$ | azanediide ion<br>hydronitride ion<br>azanediyl dianion<br>imide ion (trivial name) |
| $SH^-$ | sulfanide ion<br>hydrosulfide ion |
| $HO^-$ | hydroxide ion |
| $AlH_4^-$ | tetrahydroaluminate(1–) ion |
| $(CH_3)_2P^-$ | dimethylphosphanide ion<br>dimethylphosphide ion |

## 9.5 Polynuclear Hydrides and Their Derivatives

If a molecular hydride contains more than one central element, the skeletal atoms must form a chain, a ring, or a combination of either or both that may be homoatomic or heteroatomic. Such species may be named by general methods in use for naming inorganic chains and rings, as well as by methods more akin to organic nomenclature. *See* Chapter 10 for the detailed discussion of naming chains and rings. Trivial names, however, are in common use for a few polynuclear hydrides (e.g., hydrazine for $N_2H_4$ and hydrogen peroxide for $H_2O_2$). Some of these names are suitable as the basis for substitutive names derived according to the principles outlined in Sections 9.2–9.4 for mononuclear species.

Some of the examples that follow include names based on the principles discussed in Chapter 10.

Examples:

| | |
|---|---|
| $H_2NNH_3^+$ | hydrazinium ion<br>diazanium ion |
| $H_3NNH_3^{2+}$ | 1,2-hydrazinediium ion<br>1,2-diazanediium ion |
| $C_2H_4^{2+}$ | ethanebis(ylium) ion<br>ethanediyl dication |
| $H_2NNH^-$ | hydrazide ion<br>diazanide ion<br>trihydrodinitride ion |
| $HNNH^{2-}$ | 1,2-hydrazinediide ion<br>hydrazide(2–) ion<br>1,2-diazanediide ion<br>1,2-dihydrodinitride(2–) ion |
| $HO_2^-$ | hydroperoxide ion<br>hydrodioxide(1–) ion |
| $HF_2^-$ | hydrodifluoride ion |
| $H_2I_3^-$ | dihydrotriiodide(1–) ion |

## 9.6 Miscellaneous Hydrogen-Containing Inorganic Parent Compounds

Trivial and semisystematic names are used for some hydrogen-containing inorganic compounds, many of which are related to carbonic acid. The most common ones are listed in Table A.IX. These names are used in naming derivatives by the principles of substitutive nomenclature. Many of these compounds may also be named as functional replacement derivatives of oxo acids (*see* Sections 8.2.4 and 8.2.6).

# Chapter 10
# Chains and Rings

## 10.1 Introduction

Inorganic chemists have been increasingly concerned with chains and rings during the last half of the 20th century. The established nomenclature for such compounds was developed primarily for compounds with carbon or with carbon and heteroatoms in their skeletons and was based on organic nomenclature principles. The existing procedures must be expanded, and new ones must be developed to encompass the wide variety of homo- and heteroatomic chains and rings being created. Although a general procedure has not yet emerged, sufficient progress is being made that the great bulk of known compounds can be handled. CNIC is actively working on a document that will codify current proposals (*56*).

Naming rings and chains requires that skeletal atoms be distinguished from side-group atoms. Normally, it is obvious which atoms are skeletal in homoatomic chains and rings and heteroatomic rings, but the same is not always true for heteroatomic chains. In general, atoms that form only one bond (such as the halogens in the –I oxidation state and hydrogen) are considered to be terminal atoms, not additional atoms in the chain. Furthermore, some systems for naming chains treat atoms that might appear to be skeletal as parts of side groups (for example, *see* Section 10.3.1).

For the purposes of this chapter, some bonds that are not double bonds in classical Lewis electron-dot formulas will be written as double bonds. These are generally bonds that are shorter than expected for single bonds between the atoms in question. The decrease in bond length is usually attributed to $\pi$ bonding, and such bonding can be crudely represented by resonance among canonical formulas containing some atoms with more outer-shell electrons than the normal octet of electrons. A single canonical formula of this sort will be used herein as the basis for naming species with such bonds.

## 10.2 Homoatomic Chains

### 10.2.1 Unbranched Chains

Homoatomic chains that do not have side groups are homoatomic species, and their nomenclature is discussed in Section 2.3. Those that have side groups may be named as polynuclear hydrides or derivatives of polynuclear hydrides. The number of skeletal atoms in the chain is indicated by placing the appropriate type-1 numerical prefix (*see* Table A.II) in front of the -ane name (*see* Table A.VIII). By convention, the resulting name specifies the parent hydride, in which the appropriate number of skeletal atoms are linked into a chain by single bonds and in which there are sufficient hydrogen atoms present to give each skeletal atom its standard bonding number (*see* Section 9.2). The lambda convention (*see* Section 9.2) is used to indicate the bonding number for any skeletal atom that does not exhibit its standard bonding number.

The name of a hydride having an unbranched chain of identical noncarbon atoms with fewer hydrogen atoms than are required to satisfy the bonding numbers, standard (implied) or nonstandard (indicated by $\lambda^n$), is derived from the name of the corresponding saturated chain by changing the suffix -ane to -ene (double bond) or -yne (triple bond) and adding any necessary locants and multipliers. The parent hydride name may then be used as the basis for naming derivatives according to the principles of substitutive nomenclature. The chain of skeletal atoms is numbered by arabic numerals from one terminal atom to the other. The terminal atom to be numbered 1 is selected so that unsaturation or higher bond number has preference to substitutent prefixes for the lower locant at the first difference (*see* Section 2.2.4.6). There is nothing in current rules to indicate whether unsaturation or higher bond number is to be preferred for lower locants.

Examples:

| | |
|---|---|
| $H_3GeGeH_3$ | digermane |
| $H_2PPHPH_2$ | triphosphane |
| $H_2NNHN{=}NNH_2$<br>5 4 3 2 1 | 2-pentaazene<br>(not 3-...ene; the locant 2 is lower than 3) |
| $H_3SiSiH(CH_3)SiH_2SiH_2SiH_3$<br>1 2 3 4 5 | 2-methylpentasilane<br>(not 4-methyl...; the locant 2 is lower than 4) |
| $Ph_2NN{=}NN{=}NNPh_2$<br>6 5 4 3 2 1 | 1,1,6,6-tetraphenyl-2,4-hexaazadiene |
| $Br_3SiSiBr_2SiBr_3$ | octabromotrisilane |
| $HSSH_2SH$<br>3 2 1 | $2\lambda^4$-trisulfane |
| $H_4PPH_3PHPH_4$<br>1 2 3 4 | $1\lambda^5,2\lambda^5,4\lambda^5$-tetraphosphane<br>[not $1\lambda^5,3\lambda^5,4\lambda^5$-...; the locant set 1,2,4 is preferred to 1,3,4 (*see* Section 2.2.4.6)] |

### 10.2.2 Substituent Groups

Side groups formally derived by the loss of one or more hydrogen atoms from a chain are named by the procedures of Section 9.3. The chain is numbered from one end to the other so that the lower locant is given first to the chain atom with the free valence and then following the criteria for numbering branched chains. The abbreviated forms silyl, germyl, stannyl, and plumbyl are used instead of silanyl, etc., to distinguish more easily between a group derived from a polynuclear parent hydride and two or more groups derived from the corresponding mononuclear parent hydride (e.g., disilanyl describes the $SiH_3SiH_2$ group and disilyl describes two $SiH_3$ groups).

Examples:

| | |
|---|---|
| $H_3SnSnH_2SnH_2{-}$<br>3 2 1 | 1-tristannanyl |
| $H_2PP{=}$<br>2 1 | diphosphanylidene<br>1,1-diphosphanediyl |
| $-HPPH-$<br>1 2 | 1,2-diphosphanediyl |
| $H_3SiSiH_2Si{\equiv}$<br>3 2 1 | trisilanylidyne<br>1,1,1-trisilanetriyl |

|  |  |
|---|---|
| $H_2SbSbSb{=}$ (with a bond on Sb-2; locants 3 2 1) | 2-tristibanyl-1-ylidene<br>1,1,2-tristibanetriyl |
| $-SSH_2S-$ (locants 1 2 3) | 2λ⁴-trisulfane-1,3-diyl |

All indications are that CNOC will recommend in the next edition of the IUPAC Organic Rules that the locants be placed within the name (for example, tristannan-1-yl).

### 10.2.3 Branched Chains

A compound having a branched chain of identical noncarbon atoms is named by selecting the longest unbranched chain (the principal chain) and treating the branches as side groups. The principal chain is numbered from end to end so as to give the lower number at the first difference in side-group locants (*see* Section 2.2.4.6).

Examples:

|  |  |
|---|---|
| $H_3SiSiH_2SiH(SiH_2SiH_3)SiH_2SiH_2SiH_3$ (locants 1 2 3 4 5 6; $SiH_2SiH_3$ on Si-3) | 3-disilanylhexasilane<br>(not 4-disilanyl...; the locant 3 is lower than 4) |
| $H_3GeGeH_2GeH(GeH_3)GeH_2GeH(GeH_3)GeH_3$ (locants 6 5 4 3 2 1; $GeH_3$ on Ge-4 and Ge-2) | 2,4-digermylhexagermane<br>[not 3,5-digermyl; the locant set 2,4 is preferred to 3,5 (*see* Section 2.2.4.6)] |

### 10.2.4 Substituted Derivatives and Ions

The procedures for naming substitution products and ions derived from the hydrides of homoatomic chains are discussed in Sections 9.2 and 9.4, respectively. The names for side groups are the names used in substitutive organic chemical nomenclature (*see* Table A.V). For numbering purposes, side groups of branched chains are treated as substituents.

Examples:

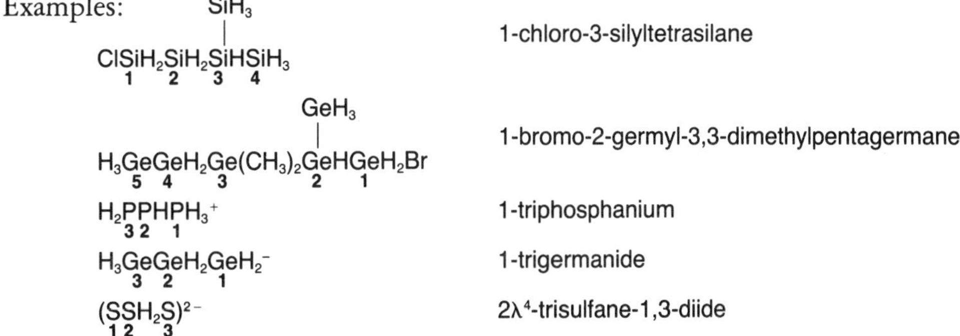


## 10.3 Heteroatomic Chains

### 10.3.1 Chains with Repeating Units

If a chain of alternating atoms is limited by identical atoms (that is, if the chain can be represented by *xyxyx*, etc.), it may be named by the following procedure. The terminal atoms of the chain (the limiting atoms) are chosen to be the identical atoms of the more electropositive element in the chain according to Figure A.1. The name is then constructed by citing successively the type-1 numerical prefix (*see* Table A.II) denoting the number of atoms of the limiting element, the "a" morpheme for that element (*see* Table A.X), the "a" morpheme for the other element in the chain, and the ending -ne, unless unsaturation dictates an ending such as -ene, -diene, or -yne instead of -ane. The final -a of an "a" morpheme is elided when it is followed by a vowel. The chain is numbered from one terminal atom to the other, with the criteria for selection of the end to be numbered 1 analogous to those for homoatomic chains (*see* Section 10.2.1). Lower locants are assigned to double and then to triple bonds if there is still a choice.

Examples:

| | |
|---|---|
| $H_3SiSeSiH_3$ | disilaselenane |
| $H_2PNHPHNHPH_2$ | triphosphazane |
| $H_3SiGeH_2SiH_2GeH_2SiH_3$ (locants 1 2 3) | 1,3-disilyldigermasilane |
| $H_3SiOSiH_2OSiHOSiH_2OSiH_3$ with $OSiH_2OSiH_3$ on atom 5 (locants 1 2 3 4 5 6 7 8 9) | 5-(disiloxanyloxy)pentasiloxane |
| $H_3CN{=}CHN{=}CH_2$ (locants 5 4 3 2 1) | 1,3-tricarbazadiene |

A provisional procedure for naming an unbranched chain of repeating units, each consisting of two or more atoms (*xyxyxy* or *xyzxyzxyz*, etc.), was proposed in the 1979 IUPAC Organic Rules (*34j*). The unitalicized prefix catena- was cited first, followed by a type-1 numerical affix denoting the number of repeating units. Then the "a" morphemes for the elements whose atoms constitute the backbone of the repeat unit were cited in parentheses in a specified order, and the name was completed with a suffix such as -ane, -ene, or -yne [for example, catenatri(phosphazane) for $H_2NPHNHPHNHPH_2$]. Names formed in this way may still be encountered in the literature; however, they are not included in current CNIC or CNOC recommendations, although such a method is still used in naming cyclic $(xy)_n$ systems (*see* Section 10.5.1).

### 10.3.2 Substituent Groups

Side groups formally derived from heteroatomic chain hydrides are named in the same way as those derived from homoatomic chain hydrides (*see* Section 10.2.2). The chain is numbered from one end to the other so that the lower locant is given first to the chain atom with the free valence and then following the criteria for numbering unbranched chains.

Examples:

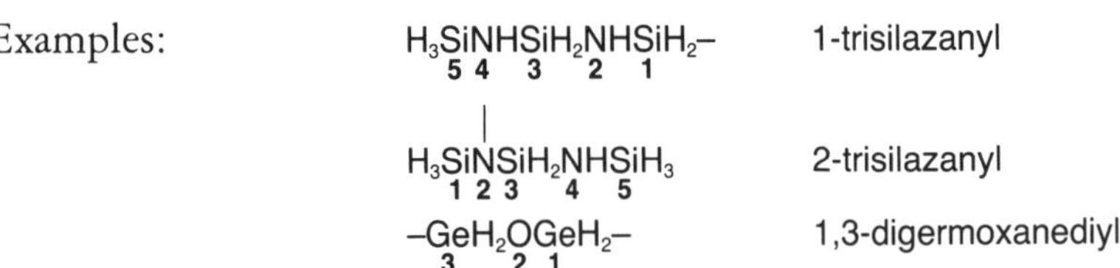


### 10.3.3 Substituted Derivatives and Ions

The general principles used for derivatives and ions of homoatomic chain hydrides (*see* Section 10.2.4) are also used for heteroatomic chain hydrides.

Examples:

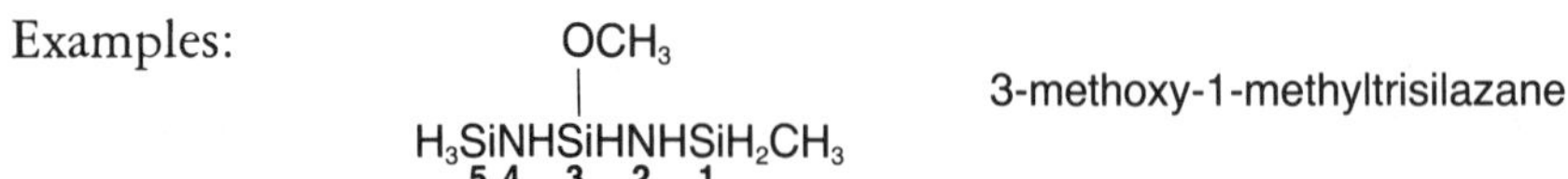

| | |
|---|---|
| $H_3SiSSiH_2SH$ | mercaptodisilathiane<br>disilathianethiol |
| $(H_3\underset{5}{Si}\underset{4}{N}\underset{3}{Si}H_2\underset{2}{N}\underset{1}{Si}H_3)^{2-}$ | 2,4-trisilazanediide ion |
| $(H_3\underset{1}{Si}\underset{2}{N}H_2\underset{3}{Si}H_2\underset{4}{N}H_2\underset{5}{Si}H_3)^{2+}$ | 2,4-trisilazanediium ion |

### 10.3.4 Chains with Random Distribution of Backbone Atoms

When the distribution of atoms of different elements in a chain is not regular, replacement nomenclature may be used for the construction of names. This procedure is discussed in Section 10.6.

## 10.4 Monocyclic Homoatomic Rings

### 10.4.1 Saturated Rings

A saturated parent ring is a ring of atoms singly bonded to each other and having standard (implied) or nonstandard (indicated by $\lambda^n$) bonding numbers (*see* Table A.X for standard bond numbers) satisfied by virtue of attachment of the appropriate number of hydrogen atoms. A compound consisting of a single saturated ring of identical noncarbon atoms may be named by adding the unitalicized prefix cyclo- to the name of the saturated unbranched chain containing the same number of identical noncarbon atoms. The ring is numbered to give the lowest possible locants to substituents (*see* Section 2.2.4.6).

Examples:

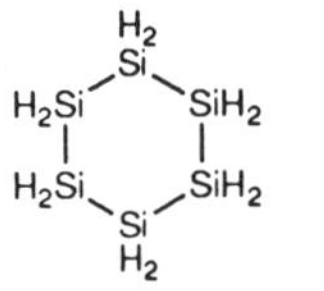

cyclohexasilane

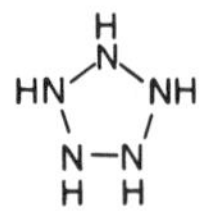

cyclopentaazane

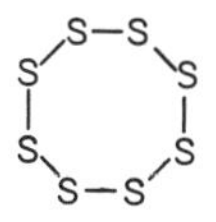

cyclooctasulfane

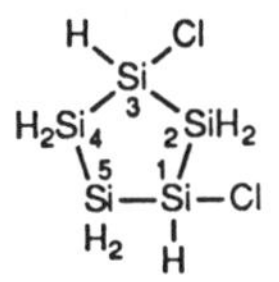

1,3-dichlorocyclopentasilane

### 10.4.2 Unsaturated Rings

The name of a parent compound consisting of a single ring of identical noncarbon atoms that contains one or more multiple bonds may be derived from the name of the corresponding saturated ring by changing the suffix -ane to -ene, -adiene, etc., and adding any necessary locants and multipliers. The origin and direction of enumeration are chosen so that the lowest locants are given to unsaturation with the same preferences that are used in Section 10.2.1.

Example:

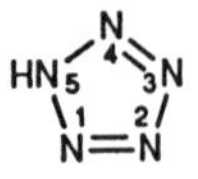

1,3-cyclopentaazadiene

### 10.4.3 Substituent Groups

A side group formally corresponding to a ring of identical noncarbon atoms with one or more free valences in place of hydrogen atoms may be named by adding the appropriate suffix -yl, or -ylidene, or -diyl, etc. (*see* Section 9.3) to the name of the parent ring (*see* Sections 10.4.1 and 10.4.2), with elision of the final "e" of the name before "y". The name is preceded by the locant(s) of the atom(s) carrying the free valence(s). The ring is numbered so that preference for the lowest locants is given first to the free valence(s), followed by unsaturation and then substituents.

Examples:

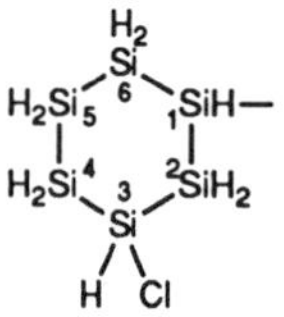

3-chloro-1-cyclohexasilanyl

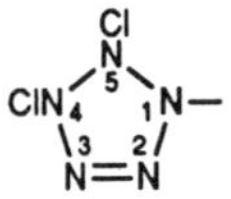

4,5-dichloro-2-cyclopentaazen-1-yl

## 10.5 Heteroatomic Rings

### 10.5.1 Rings of Repeating Units

Monocyclic systems based on repeating units linked head to tail by single bonds, with each unit consisting of atoms of two different elements, may be named as follows. The unitalicized prefix cyclo-, a type-1 numerical affix denoting the number of repeating units (*see* Table A.II), the "a" morphemes for the backbone atoms of the repeating unit (*see* Table A.X), and a suffix such as -ane, -ene, or -adiene (*see* Section 10.2.1) specifying the state of hydrogenation are cited successively. The "a" morpheme for the more electropositive element according to Figure A.1 is cited first and a terminal "a" is elided before "a" or "o". Numbering of the repeating ring atoms starts with the more electronegative atom according to Figure A.1.

Examples:

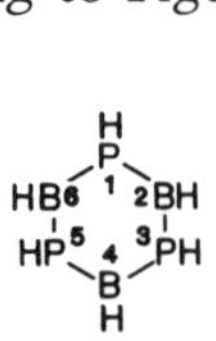

cyclotriboraphosphane

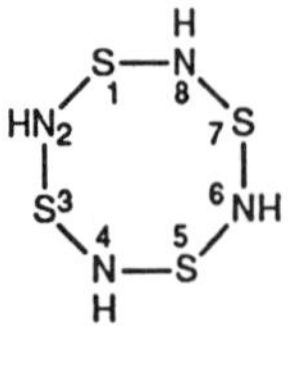

cyclotetraazathiane

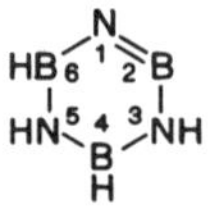

1-cyclotriborazene

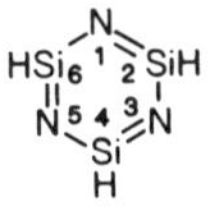

cyclotrisilazatriene

### 10.5.2 Derivatives

Side groups that formally correspond to heteroatomic ring hydrides with one or more hydrogen atoms replaced by free valences may be named in the same manner as side groups derived from homoatomic ring hydrides (*see* Section 10.4.3). All substituents are cited as prefixes in alphabetical order. If there is a choice for numbering, the ring is numbered to give the lowest locants to the substitutents at the first difference (*see* Section 2.2.4.6).

Examples:

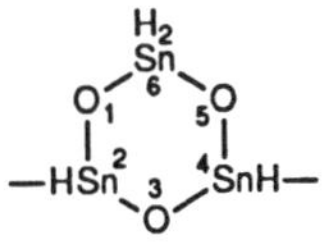

2,4-cyclotristannoxanediyl
[not 1,3-diyl...; O (oxa) is more electronegative than Sn (stanna) according to Figure A.1]

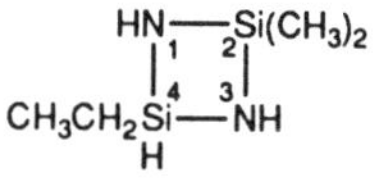

4-ethyl-2,2-dimethylcyclodisilazane
[not 3-ethyl-1,1-dimethyl...; N (aza) is more electronegative than Si (sila) according to Figure A.1 and not 2-ethyl-4,4-dimethyl... because the rearranged locant set 2,2,4 is preferred to 2,4,4 (*see* Section 2.2.4.6)]

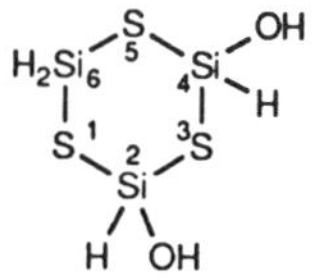

2,4-cyclotrisilathianediol
2,4-dihydroxycyclotrisilathiane
[not 1,3-...diol or 1,3-dihydroxy...; S (thia) is more electronegative than Si (sila) according to Figure A.1]

### 10.5.3 Extended Hantzsch–Widman Nomenclature

***10.5.3.1 Parent Rings.*** The Hantzsch–Widman system, which was developed for naming organic heteromonocyclic compounds, has gradually been refined (*34k, 58*) since its introduction. The general approach is to indicate the heteroatoms by "a" morphemes (*see* Table A.X) and the size of the ring by suffixes, one set for saturated rings and another for rings with the maximum number of noncumulative double bonds (*see* Table A.XI). Use of the Hantzsch–Widman system for homoatomic and heteroatomic monocyclic rings that do not contain ring carbon atoms simply requires that all the ring atoms be specified. Consequently, it is a straightforward application of the system. The name may be formed by adding the appropriate morpheme for each heteroatom, together with the appropriate type-1 numerical prefix (*see* Table A.II) in the order of appearance of the heteroatoms in Table A.X, to the appropriate suffix. The final -a of an "a" morpheme or a type-1 numerical prefix is elided before a following vowel. The numbering in heteroatomic rings proceeds from the atom that is the highest in Table A.X in the direction that gives the lower locants to the remaining atoms in the ring that are the highest in Table A.X. If there is still a choice in numbering, it is made so as to give substituents the lower numbers (*see* Section 10.2.1). The numerical locants giving the positions of the ring atoms are cited before the first heteroatom morpheme (and its associated numerical prefix, if any) and in the same order as the heteroatom morphemes to which they refer.

Examples:

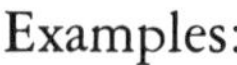

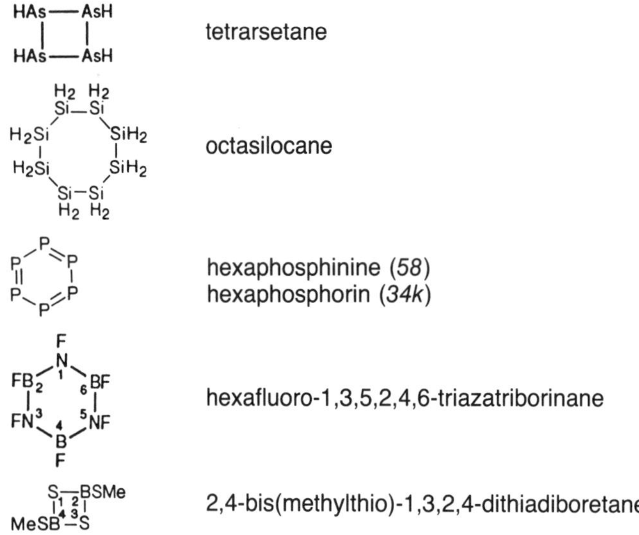


*10.5.3.2 **Indicated Hydrogen.*** The presence of at least one hydrogen atom on a ring atom joined to adjacent ring atoms by single bonds in a ring structure that otherwise contains the maximum number of noncumulative double bonds is indicated by the locant for the hydrogen atom followed by the symbol *H* placed in front of the name. Indicated hydrogen takes precedence after heteroatoms for low locants.

Examples:

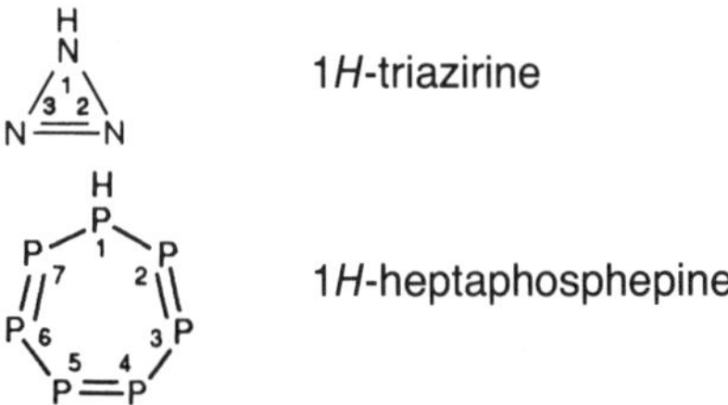


*10.5.3.3 **Partially Unsaturated Rings.*** A heteroatomic ring containing fewer than the maximum number of noncumulative double bonds may be named by adding the prefixes dihydro-, tetrahydro-, etc., to a parent name that indicates the presence of the maximum number of noncumulative double bonds or by adding the prefixes didehydro-, tetradehydro-, etc., to a parent name that indicates a fully saturated structure (*see* Section 10.5.3.1). The first method is usually preferred. A method involving the use of $\Delta$ to denote a double bond (*34l*), which is no longer recommended, may be encountered in the literature.

Examples:

1,2,3,4-tetrahydrohexazine
didehydrohexazinane

dihydrotetrarsete
didehydrotetrarsetane

### 10.5.4 Ring Atoms with Nonstandard Bonding Numbers

A nonstandard bonding number for a neutral ring atom in a parent hydride is indicated by the lambda convention (57) (*see* Section 9.2). The $\lambda^n$ symbol is cited immediately after the locant for the atom with the nonstandard bonding number. Should the locant for such an atom not be expressed in the name of the parent hydride, the locant, if necessary, and the $\lambda^n$ symbol are cited in front of the name of the parent hydride but after any indicated hydrogen. Numbering of parent hydrides with ring atoms having nonstandard bonding numbers follows the procedure for ring atoms with standard bonding numbers (*see* Section 10.5.3.1). When a further choice is needed, the order of preference for assignment of lowest locants follows the order of decreasing numerical value of the bonding number.

Examples:

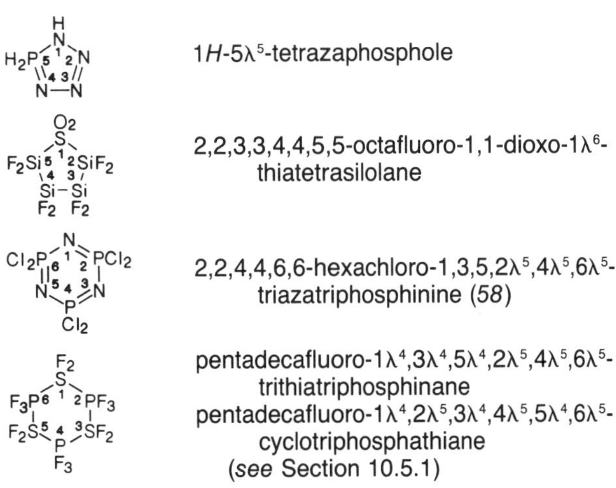

1*H*-5$\lambda^5$-tetrazaphosphole

2,2,3,3,4,4,5,5-octafluoro-1,1-dioxo-1$\lambda^6$-thiatetrasilolane

2,2,4,4,6,6-hexachloro-1,3,5,2$\lambda^5$,4$\lambda^5$,6$\lambda^5$-triazatriphosphinine (*58*)

pentadecafluoro-1$\lambda^4$,3$\lambda^4$,5$\lambda^4$,2$\lambda^5$,4$\lambda^5$,6$\lambda^5$-trithiatriphosphinane
pentadecafluoro-1$\lambda^4$,2$\lambda^5$,3$\lambda^4$,4$\lambda^5$,5$\lambda^4$,6$\lambda^5$-cyclotriphosphathiane (*see* Section 10.5.1)

### 10.5.5 Cumulative Double Bonds

Cumulative double bonds are present when more than one skeletal double bond is associated with a single ring atom. Such an atom, in a cyclic parent hydride that otherwise has the maximum number of noncumulative double bonds, is indicated by the symbol $\delta^c$, in which $c$ is an arabic number representing the number of double bonds (59). This symbol is cited immediately after the locant for the ring atom in the name for the parent hydride and follows the $\lambda^n$ symbol (*see* Section 10.5.4), if present. If the locant for the ring atom is not expressed in the name of the parent, the locant and the $\delta^c$ symbol are cited in front of the name of the parent hydride but after any indicated hydrogen. When there is still a choice for numbering ring atoms, a ring atom with cumulative skeletal double bonds attached to it is preferred to the same ring atom with the same bonding number but without cumulative skeletal double bonds.

Examples:

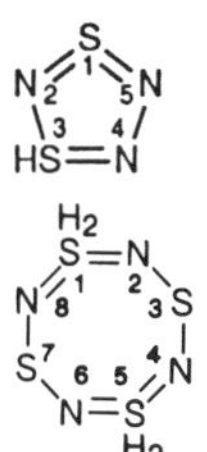

1$\lambda^4\delta^2$,3$\lambda^4$,2,4,5-dithiatriazole

1$\lambda^6\delta^2$,3,5$\lambda^6\delta^2$,7,2,4,6,8-tetrathiatetrazocine
1$\lambda^6\delta^2$,5$\lambda^6\delta^2$-cyclotetrazathiatetraene

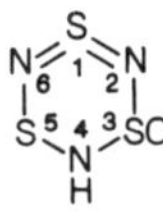

3,4-dihydro-3-oxo-1λ$^4$δ$^2$,3λ$^4$,5,2,4,6-trithiatriazine
(The parent ring structure has a double bond between positions 3 and 4, so "3,4-dihydro" is required to indicate that the double bond has been saturated.)
3-oxo-1λ$^4$δ$^2$,3λ$^4$-cyclotriazathiadiene
(*see* Section 10.5.1)

### 10.5.6 Problems with Boron Compounds

Apparently the intent of the original proposal regarding $\lambda^n$ was that coordinate covalent bonds be counted in determining bond number. The wording, however, confused partial ionic character of such a bond with the charge on a distinct ion. It seems reasonable to conclude that when coordinate covalent bonds are present in boron–nitrogen compounds, boron and nitrogen may both be considered to have a bonding number of four. Names for a variety of neutral boron–nitrogen compounds may then be constructed.

Examples:

1,3,5,2,4,6-triazatriborinine
cyclotriborazatriene

1,3,5,2,4,6-triazatriborinane
cyclotriborazane
(traditionally borazine)

1λ$^4$,3λ$^4$,5λ$^4$,2λ$^4$,4λ$^4$,6λ$^4$-triazatriborinane

### 10.5.7 Ions

The addition of a proton to a ring atom is indicated by changing the final -e in the suffix of the name for the parent ring to -ium. If the cation may be considered to be derived formally by the loss of one or more electrons from a neutral parent compound, it may be named by adding the term "radical cation($n+$)", in which $n+$ represents the charge on the cation, to the name of the parent compound.

Examples:

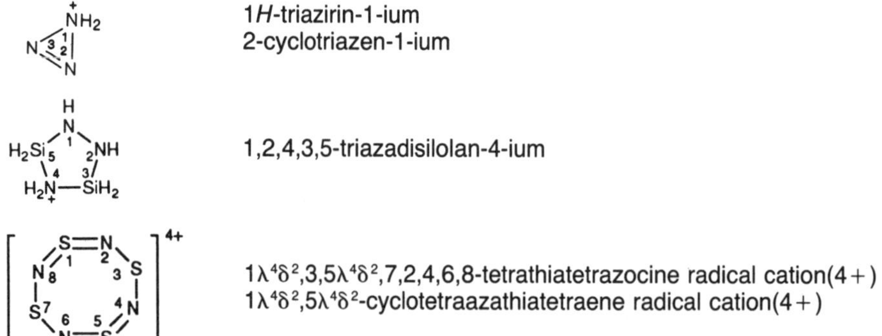

1*H*-triazirin-1-ium
2-cyclotriazen-1-ium

1,2,4,3,5-triazadisilolan-4-ium

1λ$^4$δ$^2$,3,5λ$^4$δ$^2$,7,2,4,6,8-tetrathiatetrazocine radical cation(4+)
1λ$^4$δ$^2$,5λ$^4$δ$^2$-cyclotetraazathiatetraene radical cation(4+)

The loss of a proton from an organic acid or hydroxy compound is usually indicated by the suffix -ate. If the anion may be considered to be derived formally by the addition of one or more electrons to a neutral compound, it may be named by adding the term "radical anion($n$–)", in which $n$– represents the charge on the anion, to the name of the parent compound. Anions formed by the loss of one or more protons from a parent ring may be named by the method presented in Section 9.4.

Examples:

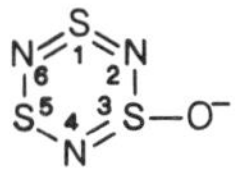

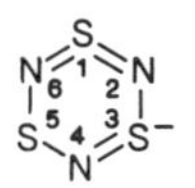

$1\lambda^4\delta^2,3\lambda^4$,5,2,4,6-trithiatriazin-3-olate
($\lambda^4\delta^2$S is numbered first, then the other $\lambda^4$S)
$1\lambda^4\delta^2,3\lambda^4$-cyclotriazathiatrien-3-olate

$1\lambda^4\delta^2,3\lambda^4$,5,2,4,6-trithiatriazin-3-ide
($\lambda^4\delta^2$S is numbered first, then the other $\lambda^4$S)
$1\lambda^4\delta^2,3\lambda^4$-cyclotriazathiatrien-3-ide

## 10.6 Replacement Nomenclature

The procedure known as replacement nomenclature (*34m*) is a method for naming cyclic and acyclic hydrocarbons in which some of the carbon atoms with attached hydrogen atoms are conceptually replaced by heteroatoms (with attached hydrogen atoms when appropriate). Although there are no published guidelines, this method could be used in a generalized form for compounds that can be viewed as atomic chain or ring hydrides containing one or more different skeletal atoms. In such cases the name for the parent homoatomic species could be based on the element supplying the largest number of skeletal atoms. In the examples following the next paragraph, the parent homoatomic chain or ring is based on the more (most) electropositive element according to Figure A.1 when two (or more) elements each supply the same (larger or largest) number of skeletal atoms.

The name is constructed by prefixing "a" morphemes (*see* Table A.X) with appropriate locants to the name for the parent homoatomic hydride. If more than one "a" morpheme is used, the order of seniority proceeds from the top of Table A.X downward. Derivatives may be named in the same way that derivatives of homoatomic species are named (*see* Sections 10.2 and 10.4).

Examples:

$H_3CNHSCH_2NHSCH_2NHSCH_3$ (locants 10 9 8 7 6 5 4 3 2 1)
2,5,8-trithia-3,6,9-triazadecane
[not 3,6,9-trithia-2,5,8-triaza...; S (thia) precedes N (aza) in Table A.X]

$H_3SiSiH_2GeH_2SiH_2SiH_3$ (locants 5 4 3 2 1)
3-germapentasilane

$H_3CSiH_2NHNHCH_2SiH_2SSiH_3$ (locants 1 2 3 4 5 6 7 8)
7-thia-3,4-diaza-1,5-dicarbaoctasilane
[not 2-thia-5,6-diaza-4,8-dicarba...; the locant set 1,3,4,5,7 is preferred to 2,4,5,6,8 (*see* Section 2.2.4.6)]

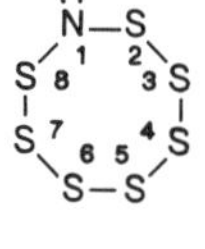

1-azacyclooctathiane

$H_3CSCH{=}CHSCH{=}CHSCH_3$ (locants 1 2 3 4 5 6 7 8 9)
2,5,8-trithia-3,6-nonadiene

$H_2NCH_2NHN{=}CHN{=}NCH_2NH_2$ (locants 9 8 7 6 5 4 3 2 1)
2,5,8-tricarba-3,5-nonaazadiene
[not ...4,6-nonaazadiene; the locant set 3,5 is preferred to 4,6 (*see* Section 2.2.4.6)]

1,4,7,10-tetraoxa-2,5,8,11-tetraazacyclo-dodecasilane

1,3,5,7,9-pentaaza-2-phosphabicyclo[5.3.1]-undecasilane

Skeletal atoms that are formally cationic due to the addition of $H^+$ may be indicated by the use of morphemes that are usually derived from the "a" morphemes by replacing the final -a with -onia. These modified morphemes are cited after the normal morpheme for the element in question, although when there is a choice the site for the charge is given the lower locant.

Examples:

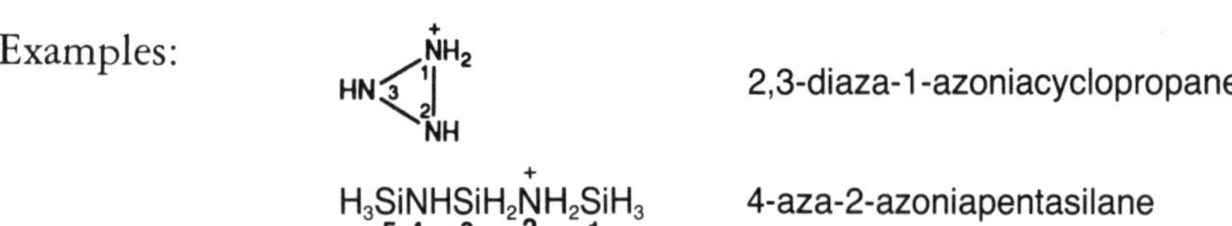

2,3-diaza-1-azoniacyclopropane

4-aza-2-azoniapentasilane

The 1979 IUPAC Organic Rules tentatively provide for a morpheme derived from "bora" by replacing the final -a with -ata (i.e., "borata") to indicate skeletal boron atoms that can be considered anionic by virtue of the formal addition of $H^-$ to them (*34n*).

Example:

$(CH_3)_2PBHP(CH_3)\bar{B}H_2P(CH_3)_2$ (locants 5 4 3 2 1) 1,1,3,5,5-pentamethyl-4-bora-2-boratapentaphosphane

Although this method was developed for boron hydrides, it has also been used for phosphorus hydrides. It has, however, not yet been officially adopted and is still being studied.

# Chapter 11
# Boron Compounds

## 11.1 Introduction

The element boron has fewer electrons in its outer shell than any other nonmetal except hydrogen or helium. In addition to its involvement in traditional types of compounds, boron is involved in bonding and structural patterns that are unique in many ways. The traditional compounds are "electron-precise"; that is, they have coplanar tricoordinate or tetrahedral structures that can be described in terms of traditional electron-pair bonds. However, boron also forms catenated molecular and anionic hydride networks featuring triangulated clusters of boron atoms, in contrast to the usual chain and ring structures formed by other nonmetals. These structures appear consistent with the structure found in elemental boron, metal borides, and other compounds with high boron content, such as boron carbides and nitrides.

Naming the full range of boron-containing ions and compounds requires the use of a variety of nomenclature systems and techniques. Homoatomic and binary species and electron-precise (traditional) compounds may be named by the principles of additive or substitutive nomenclature already discussed. It is necessary, however, to extend and modify those systems in order to adapt them to the special characteristics of the polyboron clusters. Combinations of principles from different systems have been proposed because no one system has proved entirely adequate by itself.

## 11.2 Standard Names

Homoatomic and binary boron-containing species and electron-precise boron compounds may be named according to the procedures in this book appropriate to the type of species being named.

Examples of homoatomic species (*see* Chapter 3):

| | |
|---|---|
| $B^{3+}$ | boron(3+) ion<br>boron(III) cation |
| $[B_2]^+$ | diboron(1+) ion |
| $[B_6]^{2-}$ | hexaboride(2−) ion |
| $B_6$ | hexaboron |

Examples of stoichiometric names (*see* Chapter 4):

| | |
|---|---|
| CaB | calcium boride |
| $Pt_3B_2$ | triplatinum diboride |

| | |
|---|---|
| $B_3Si$ | triboron silicide |
| $B_{11}C_2$ | undecaboron dicarbide |
| CeCoB | cerium cobalt boride |
| $Lu_2ReB_6$ | dilutetium rhenium hexaboride |
| $B_2NaFO_3$ | diboron sodium fluoride trioxide |

Examples of additive names (*see* Chapter 5):

| | |
|---|---|
| $[BH_4]^-$ | tetrahydroborate(1 −) ion[a] |
| $[BF_4]^-$ | tetrafluoroborate(1 −) ion |
| $B(CH_3)_3$ | trimethylboron |

[a]It is traditional to use "hydro" rather than "hydrido" for hydrogen as an anionic ligand in naming boron coordination compounds.

Examples of traditional oxo acid and related names (*see* Sections 8.2 and 8.4):

| | |
|---|---|
| $H_3BO_3$ | boric acid (*see* Section 8.2.3) |
| $HBO_2$ | metaboric acid (*see* Section 8.2.3) |
| $ClB(OH)_2$ | chloroboric acid<br>borochloridic acid |
| $H_2B(SeH)$ | selenoborinic acid<br>borinoselenoic acid |
| $HB(NH_2)(OH)$ | amidoboronic acid<br>boronamidic acid |
| $HB_3O_5$ | triboric acid<br>hydrogen pentaoxotriborate |
| HB(OH)OBH(OH) | diboronic acid<br>dihydrogen μ-oxo-dioxodiboronate |
| [ring structure: HO—B, O, B—OH, O, O, B, OH] | trihydrogen *cyclo*-tri-μ-oxo-trioxotriborate<br>*cyclo*-triboric acid |
| $(CH_3O)_3B$ | trimethyl borate |
| $CH_3BO_2^{2-}$ | methylboronate(2 −) ion |

Examples of substitutive names (*see* Chapter 9):

| | |
|---|---|
| $BH_3$ | borane |
| $BHCl_2$ | dichloroborane |
| $BCl_2^+$ | dichloroboranylium |

Examples of acyclic names (*see* Sections 10.3 and 10.6):

| | |
|---|---|
| $H_3CBHCH_2BHCH_2BHCH_3$ (7 6 5 4 3 2 1) | 2,4,6-triboraheptane<br>bis[(methylboryl)methyl]borane (substitutive) |
| $H_2BPHBHPHBH_2$ (5 4 3 2 1) | 2,4-diphosphapentaborane<br>triboraphosphane<br>bis(borylphosphino)borane (substitutive)<br>2,4-diphospha-1,3,5-triborapentane (organic replacement) |

Examples of heterocyclic names (*see* Sections 10.5 and 10.6):

| | |
|---|---|
| [ring structure: H, B 1, $H_2C$ 5, 2 $CH_2$, C 4, 3 C, $H_2$ $H_2$] | borolane<br>boracyclopentane |

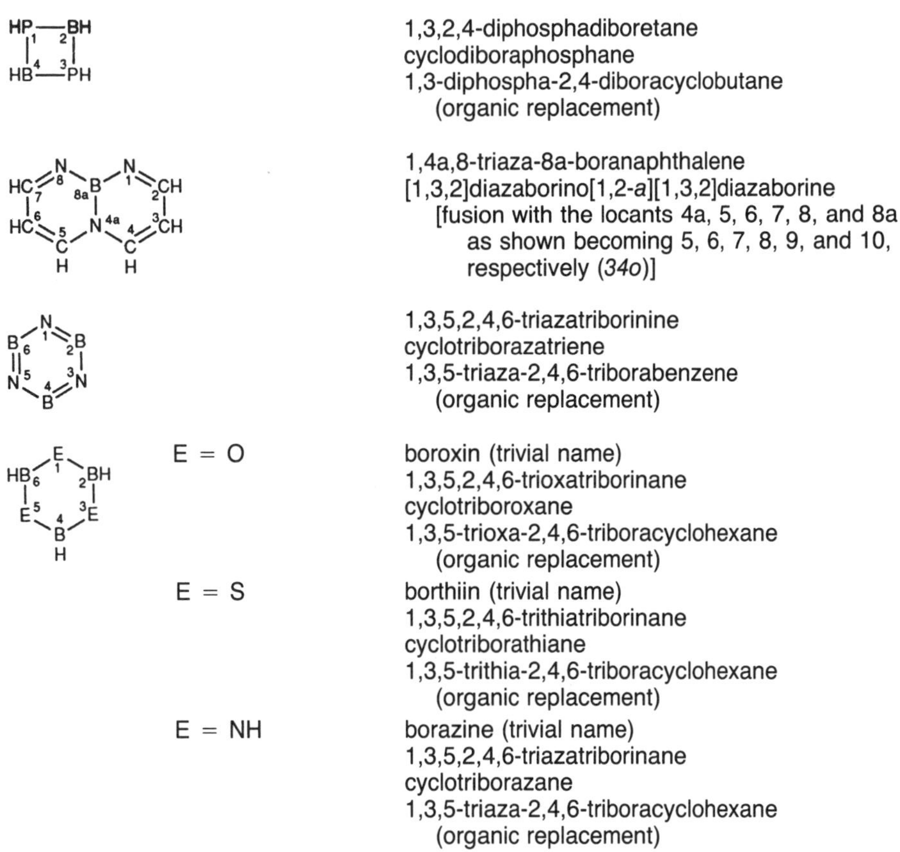

|  |  |  |
|---|---|---|
| | | 1,3,2,4-diphosphadiboretane<br>cyclodiboraphosphane<br>1,3-diphospha-2,4-diboracyclobutane<br>(organic replacement) |
| | | 1,4a,8-triaza-8a-boranaphthalene<br>[1,3,2]diazaborino[1,2-*a*][1,3,2]diazaborine<br>[fusion with the locants 4a, 5, 6, 7, 8, and 8a as shown becoming 5, 6, 7, 8, 9, and 10, respectively (*34o*)] |
| | | 1,3,5,2,4,6-triazatriborinine<br>cyclotriborazatriene<br>1,3,5-triaza-2,4,6-triborabenzene<br>(organic replacement) |
| | E = O | boroxin (trivial name)<br>1,3,5,2,4,6-trioxatriborinane<br>cyclotriboroxane<br>1,3,5-trioxa-2,4,6-triboracyclohexane<br>(organic replacement) |
| | E = S | borthiin (trivial name)<br>1,3,5,2,4,6-trithiatriborinane<br>cyclotriborathiane<br>1,3,5-trithia-2,4,6-triboracyclohexane<br>(organic replacement) |
| | E = NH | borazine (trivial name)<br>1,3,5,2,4,6-triazatriborinane<br>cyclotriborazane<br>1,3,5-triaza-2,4,6-triboracyclohexane<br>(organic replacement) |

Examples of names for substituent groups based on boron (*see* Section 9.3):

| | |
|---|---|
| $H_2B$– | boryl |
| HB= | borylidene[a] |
| –BH– | boranediyl |
| B≡ | borylidyne[a] |
| –B– (with a third bond drawn upward) | boranetriyl |
| –B= | boranylylidene[a] |
| $(HO)_2B$– | borono |

[a]Refer to Section 9.3 for the meaning of the suffix.

The names boryl, borylidene, and borylidyne, which are contractions for boranyl, etc., are used so that the numerical prefixes bis-, tris-, etc., are not needed to denote two or more such groups in order to avoid ambiguity with names of polyborane groups such as diboranyl. The name borylene, which was used to denote –BH– or HB=, is no longer recommended.

## 11.3 Polyboron and Heteropolyboron Hydrides

### 11.3.1 Stoichiometric Names

Polyboron hydrides may be named without regard to structure by adding a type-1 numerical prefix (*see* Table A.II) to give the number of boron atoms in the hydride to the stem bor-, followed by an ending appropriate to the species being named. If the polyboron hydride is neutral, the ending is -ane, to which is attached a parenthetical arabic number denoting the number of hydrogen atoms in the hydride. If the polyboron hydride is a cation, the ending is -on followed by the parenthetical Ewens–Bassett number (*see* Section 2.3.3), and the number of hydrogen atoms is indicated by a prefix consisting of the appropriate type-1 numerical prefix followed by hydro. If the polyboron hydride is an anion, the ending is -ate followed by the Ewens–Bassett number, and the number of hydrogen atoms is indicated in the same way as for a cation. Replacement of skeletal boron atoms of a polyboron hydride by other atoms is indicated by "a" morpheme prefixes (*see* Section 10.6).

Examples:

| | |
|---|---|
| $B_2H_6$ | diborane(6) |
| $B_4F_6$ | hexafluorotetraborane(6) |
| $B_{10}H_{14}$ | decaborane(14) |
| $B_{10}C_2H_{12}$ | dicarbadodecaborane(12) |
| $[B_{10}H_{12}]^{2-}$ | dodecahydrodecaborate(2−) ion |
| $[B_{11}SH_{12}]^{+}$ | dodecahydrothiadodecaboron(1+) ion |

### 11.3.2 Substitutive Names

Structural specificity for electron-precise polyboron hydrides and their derivatives may be provided by the usual rules of organic substitutive nomenclature for chains and rings (*34i*). The use of a parenthetical arabic number to indicate the number of hydrogen atoms is usually not necessary.

Examples:

| | |
|---|---|
| $H_2B\text{–}B\text{=}BH$ (B atoms numbered 3 2 1) | 1-triborene |
| $F_2BB(BF_2)BF_2$ (central chain $F_2BBBF_2$ numbered 3 2 1, with $BF_2$ on B-2) | 2-(difluoroboryl)tetrafluorotriborane |
| $H_2BBH_2$ | diborane(4)[a] |
| cyclic $(BH)_3$ (HB bonded to two BH, which are bonded to each other) | cyclotriborane<br>triborirane |

[a]The name "diborane", which would be expected, should not be used without the parenthetical numerical suffix, because without it this name is commonly used to indicate diborane(6).

### 11.3.3 Traditional Descriptors

Structural specificity for a limited number of polyboron hydrides with skeletal structures that are triangulated polyhedrons or fragments of triangulated polyhedrons may be provided by italicized prefixes such as *closo*- and *nido*- (*24*). Efforts to systematize the characterization of polyboron hydride structures by such prefixes have encountered problems because of the increasing number of possible structural variants as the size of the polyhedron increases, especially in metallapolyboranes. Nevertheless, the employment of such prefixes for the common polyboron hydrides has proved so convenient that they are used quite generally.

The prefix *closo*- may be used to designate both structure and numbering of certain fully triangulated polyboron hydride polyhedrons that have from 4 to 12 skeletal boron atoms. The names for these species take the form *closo*-(numerical prefix)bor(ending). Their structures and numbering are shown in entries 1–5, 7, 10, 12, and 14 in Chart A.1, part A. The prefix *isocloso-* has been used in two cases to distinguish between isomeric closed polyhedral structures (*60*). The structures and numbering for these *isocloso-*polyhedrons are shown in entries 8 and 11 in Chart A.1, part A. It has also been suggested that metallaboranes with the closed polyhedral structure 12 in Chart A.1, part A be called "isocloso" rather than "closo" (*61*).

The prefix *nido-* may be used to designate both the structure and numbering for a polyhedron derived from a *closo*-polyhedron by removal of the vertex or one of the vertices with the highest skeletal connectivity, provided that, in the latter case, only one structure is created. However, because there are other definitions for "nido" based on a $B_nH_{n+4}$ stoichiometry for the neutral polyboron hydride or a skeletal electron pair count of $(n + 2)$, only polyboranes 1–4 in Chart A.1, part B, for which all definitions agree, should be described as "nido". The prefix *isonido-* has been used to designate the polyhedron derived from the closed undecapolyhedral structure 12 in Chart A.1, Part A by removal of one of its vertices with the lowest skeletal connectivity (*62*). The structure and numbering of this "isonido" polyhedron are shown in entry 5 in Chart A.1, part B.

The prefix *arachno-* may be used to designate both the structure and numbering of a polyhedron derived by the removal of the vertex or one of the vertices of highest skeletal connectivity from the open face of a "nido" structure, provided that, in the latter case, only one structure can be produced. Other definitions of arachno are based on a $B_nH_{n+6}$ stoichiometry for the neutral boron hydride or a skeletal electron pair count of $(n + 3)$. Therefore, only polyboranes 7–11 in Chart A.1, part B, for which all definitions agree, should be described as "arachno". The structure of *arachno*-octaborane(14), shown in Chart A.1, part B as structure 10, does not meet the requirement regarding generation of a single structure when one of the vertices of highest skeletal connectivity is removed from the open face of a *nido*-polyhedron. The *nido*-polyhedron, structure 6 in Chart A.1, part B, has two pairs of four-coordinate vertices in the open face that are different. Only removal of one of the vertices next to the four-coordinate cap will generate the structure for *arachno*-octaborane(14), shown in Chart A.1, part B, as structure 10. The prefix *isoarachno-* has been used to describe the nonaborane(15) isomer whose skeleton is shown as structure 12 in Chart A.1, part B and is derived from the skeleton for *nido*-decaborane(14) (structure 3 in Chart A.1, part B) by removal of a vertex of lowest skeletal connectivity from its open face (*62*).

The prefix *hypho-* has been used to describe polyboron hydrides with $(n + 4)$ skeletal pairs (*63*). Only a few examples of this class, which all can be derived from a *closo*-polyhedron by the loss of three vertices, are known. For example, the skeletal structure of $[B_5H_{12}]^{1-}$ (*64*) and $B_5H_9[P(CH_3)_3]_2$ (*65*) is

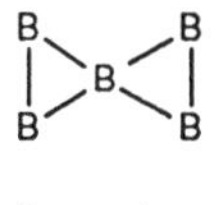

B B
B
B B

which can be derived by the removal of vertices 2, 5, and 8 from the *closo*-polyhedron shown as structure 6 in Chart A.1, part A. However, the skeletal structure reported for $B_6H_{10}[P(CH_3)_3]_2$ (*66*), which is called "hypho", is the same as that for *arachno*-$B_6H_{12}$, shown as structure 9 in Chart A.1, part B. The prefix *klado-* has been suggested for designating even more open structures for polyboron hydrides with $(n + 5)$ skeletal electron pairs (*67*), but it does not appear to have been used.

Several prefixes have been suggested for designating polyboron hydrides with ($n$) skeletal electron pairs. Because there has not been a general consensus on their use, only their published definitions are noted here.

- ▶ *precloso-* an ($n$ − 1)-vertex structure plus one capping BH vertex (*68*)
- ▶ *hypercloso-* an ($n$)-vertex cluster with less than ($n$ + 1) skeletal electron pairs and significant distortion from the idealized "closo" structure (*69*)
- ▶ *pileo-* an ($n$)-vertex cluster with ($n$) electron pairs and the exclusive geometry that clusters with such skeletal electron counts may adopt (*70*)

Two prefixes, *canasto-* and *anello-*, have been suggested as descriptors for specific open structures of certain carbaboranes (*71*). The prefix *canasto-* indicates a structure with two nontriangular faces and the general shape of a basket. The prefix *anello-* describes a circular structure consisting of one or two parallel planes of vertices arranged in such a way that the addition of two vertices, one on each side of the structure, produces a closed bipyramidal structure. These prefixes have not found much use.

## 11.3.4 The CEP Descriptor

***11.3.4.1 General Discussion.*** Structural specificity for virtually all polyboron hydrides with structures that are triangulated polyhedrons, fragments of triangulated polyhedrons, or any combination thereof may be provided by the CEP descriptor system (*53, 72*). This system is based on five-part alphanumeric descriptors for fully triangulated polyhedrons having at least one rotational symmetry axis and at least one symmetry plane (*53a*). The descriptor consists of

1. an arabic numeral giving the number of vertices in the polyhedron followed by the italic letter $v$, both enclosed in a single set of parentheses
2. the point group symbol giving the symmetry of the idealized triangulated polyhedron
3. a series of arabic numerals and symbols $v^n$, all enclosed in a single set of parentheses, giving the arrangement and type of vertices in the idealized triangulated polyhedron (*see* Section 11.3.4.4)
4. the symbol $\Delta^f$, giving the number $f$ of triangular faces in the idealized triangulated polyhedron
5. the descriptive morpheme *-closo*

The last four parts of the full descriptor are enclosed in brackets and separated from each other by hyphens. Thus, the full descriptor for the icosahedral $B_{12}$ polyhedron is $(12v)[I_h\text{-}(1v^5 5v^5 5v^5 1v^5)\text{-}\Delta^{20}$-*closo*].

Simplification of this symbolism without impairing its generality is difficult. However, for the common boron polyhedrons with 12 or fewer vertices, part 1 (the total number of vertices), which is also equal to the sum of the primary numbers in part 3 and the numerical prefix giving the number of skeletal vertices, need not be cited, and the $v^n$ symbol need not be cited when $n = 5$ or when $n = 3$ or 4 and there are no vertices in the polyhedron with $n > 5$. Hence, an abbreviated symbol for the icosahedral $B_{12}$ polyhedron would be $[I_h\text{-}(1551)\text{-}\Delta^{20}$-*closo*].

The full descriptor precedes the numerical prefix that gives the number of boron atoms in the name of a polyboron hydride. Prefixes indicating skeletal heteroatoms are cited in front of the structural descriptor. Thus, the name for one of the isomers of dicarbadodecaborane(12) would be 1,2-dicarba[$I_h$-(1551)-$\Delta^{20}$-*closo*]dodecaborane(12) (abbreviated symbolism).

***11.3.4.2 Reference Axis and Terminal Plane.*** Part 3 of the CEP descriptor requires the selection of a reference axis and a terminal plane. The discussion that follows may be facilitated by reference to Figure 11.1. The reference axis is a rotational symmetry axis of highest order lying in a symmetry plane of the polyhedron. If a further choice is needed, the reference axis should (1) have a terminal plane with the fewest polyhedral vertices or (2) lie in the reference plane (*see* Section 11.3.4.3).

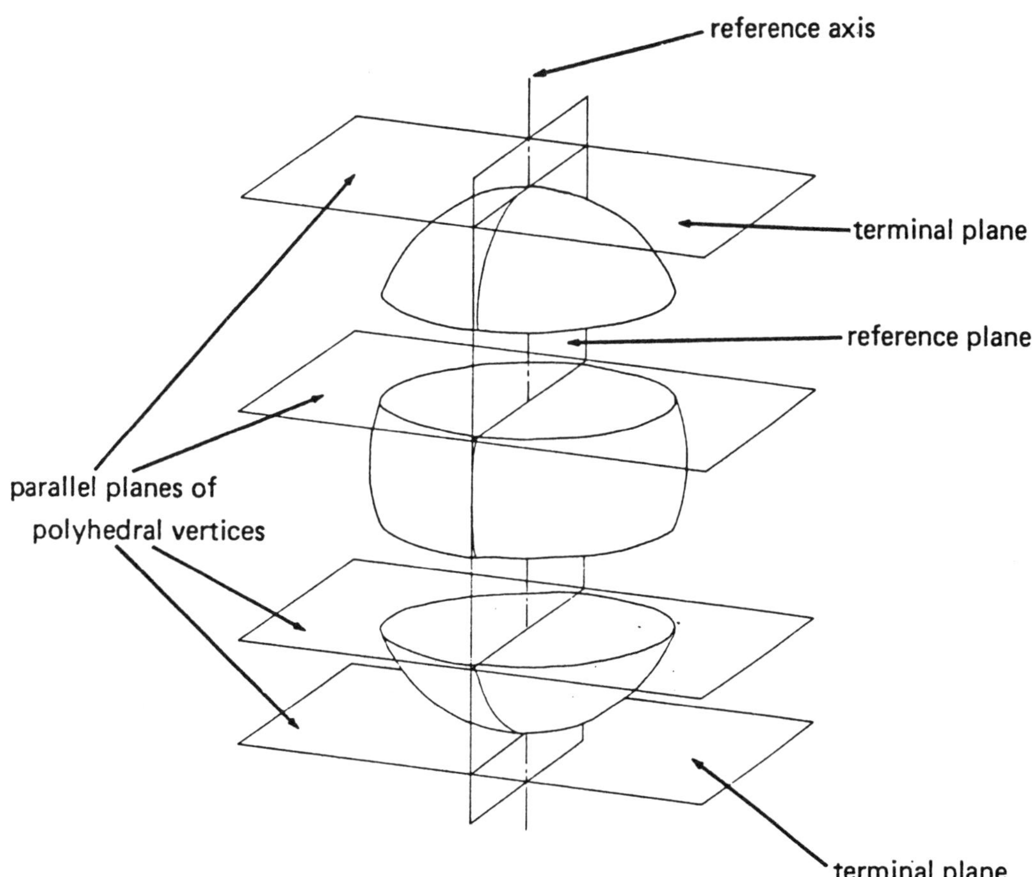


*Figure 11.1. Reference axis and symmetry planes.*

The preferred end of the reference axis is the end with the preferred terminal plane, which is the terminal plane that contains fewer polyhedral vertices. If both terminal planes contain the same number of polyhedral vertices, the preferred terminal plane is the one that contains the vertices with the lower total number of skeletal connectivities (i.e., fewer polyhedral edges associated with the vertices). If a further choice is needed, the preferred terminal plane is the terminal plane nearer to the parallel plane of symmetrically equivalent vertices preferred by applying the preceding criteria successively to pairs of parallel planes, proceeding inward from each terminal plane.

*11.3.4.3* ***Reference Plane.*** The reference plane is a symmetry plane or, if there is none, an arbitrarily defined plane containing the reference axis and at least one vertex not on the reference axis in (1) a terminal plane or (2) an adjacent parallel plane. If there is more than one such plane containing the reference axis or if selection of a reference axis is not possible by the preceding criteria, the reference plane is chosen by applying the following criteria sequentially until a choice can be made.

1. If the polyhedron has more than one symmetry plane, the reference plane is the plane that passes through the fewest vertices.
2. If the polyhedron does not have a symmetry plane, the reference plane is the plane that has the fewest perpendicular planes defined by sets of polyhedral vertices that are either symmetrically equivalent or connectively equivalent (i.e., that have the same number of associated polyhedral edges).
3. The reference plane is the plane that passes through a polyhedral vertex that is nearer to the preferred terminal plane (i.e., polyhedral vertices in the reference plane have lower locant numbers when assigned according to the numbering principles discussed in Section 11.3.4.5).
4. The reference plane is the plane that passes through polyhedral vertices with the fewest associated polyhedral edges (i.e., polyhedral vertices with the lowest skeletal connectivity in the reference plane have the lower polyhedral locant numbers when assigned according to the numbering principles discussed in Section 11.3.4.5).

*11.3.4.4* ***Part 3 of the CEP Descriptor.*** The arrangement and type of vertices in each plane perpendicular to the reference axis of the polyhedron (*see* Section 11.3.4.2) are given in part 3 of the CEP descriptor (*see* Section 11.3.4.1) by the symbol $mv^n$, in which $m$ is the number of vertices with the same skeletal connectivity in the plane and $n$ is the number of polyhedral edges associated with each vertex. The symbols are cited in the order in which the planes they define are preferred for numbering (*see* Section 11.3.4.5).

*11.3.4.5* ***Polyhedron Numbering.*** A polyhedron is oriented for numbering (*53*) by looking down the reference axis from the preferred end and rotating the polyhedron until the reference plane is vertical in order to provide a consistent reference point. If a choice is needed, the preferred orientation has a polyhedral vertex of the preferred terminal plane, or in a parallel plane nearest to it, at, or nearer, the top of the polyhedron in the reference plane.

The vertices of a polyhedral framework are numbered consecutively, clockwise or counterclockwise, starting with vertices in the preferred terminal plane and proceeding to succeeding planes of vertices working along the reference axis away from the preferred end. The vertices in each plane are numbered in the same direction as the preceding and/or following planes, beginning in each plane with a vertex at the top of the preferred orientation, in the reference plane, or with the first vertex encountered moving from the top of the preferred orientation in the direction chosen or required for numbering. A vertex at the top of the preferred orientation need not be exactly in the reference plane for chiral polyhedrons in which the reference plane is arbitrarily chosen (*53b*), but it should be very close to the reference plane.

The $D_{2d}$ dodecahedron shown in Figure 11.2 has three $C_2$ axes: "x" (passing through edges 3–6 and 4–5), "y" (passing through edges 3–5 and 4–6), and "z" (passing through edges 1–2 and 7–8). It also has two symmetry planes: "A" (passing through vertices 1, 2, 5, and 6) and "B" (passing through vertices 3, 4, 7, and 8). Only the "z" axis can be the reference axis in this polyhedron because it is the only $C_2$ axis lying in a symmetry plane of the polyhedron. The two symmetry planes "A" and "B" are equivalent, in that both contain the reference axis "z" and both pass through four polyhedral vertices. Symmetry plane "A" is the reference plane, because in the preferred orientation it passes through the polyhedral vertices of a terminal plane.

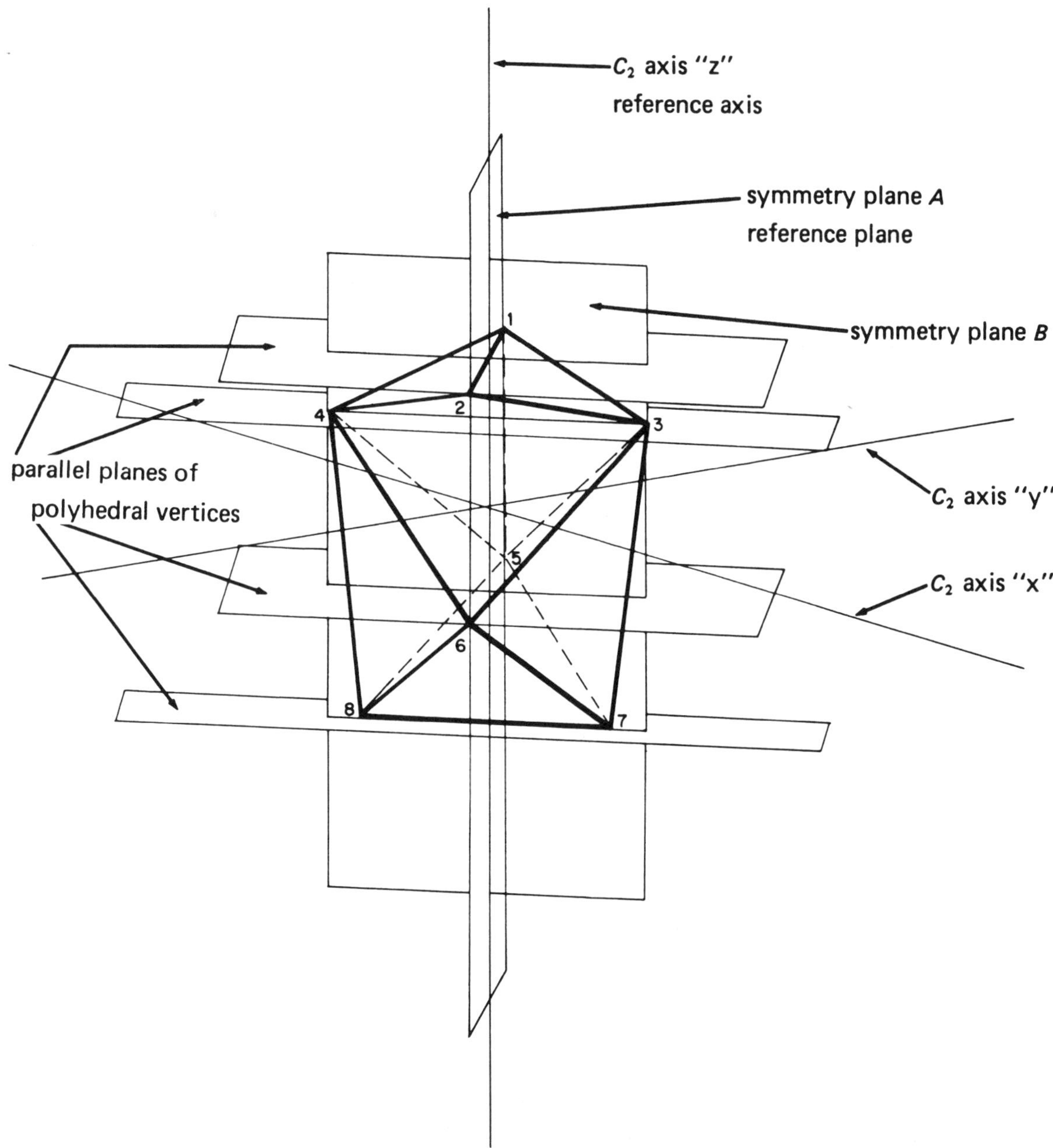


*Figure 11.2 Reference axis and symmetry planes of a dodecahedron.*

The $D_{2d}$ polyhedron is oriented for numbering by looking down the reference axis "z", in this case from either end, and rotating the structure until the reference plane "A" is vertical (i.e., perpendicular to the plane of the paper). The two vertices in the first parallel plane of vertices are numbered 1 and 2, starting at the 12 o'clock position. The two in the next parallel plane are numbered 3 and 4 in a clockwise direction from the same 12 o'clock position, and the remaining two pairs of vertices are numbered 5 and 6 and 7 and 8 in the same way.

The numbering for 24 closed polyhedrons is given in Chart A.I, Part A.

If the reference plane of the polyhedron is not a symmetry plane, the direction of numbering is determined by noting the direction from the reference plane of the directional vertex (i.e., the vertex in the directional plane nearest to the top of the projection of the reference plane) (*53b*). The directional plane is the terminal plane at the end of the reference axis opposite to the preferred terminal plane, or, if this terminal plane contains only one vertex, the directional plane is a parallel plane of polyhedral vertices adjacent to it.

The position of a heteroatom in the polyhedral structure of a heteropolyboron hydride is indicated by the lowest locant consistent with the preceding numbering rules. The presence of a heteroatom or substituents can introduce chirality. Enantiomers may be distinguished by the use of the descriptors *C* and *A* (*53b, 72c*) (*see also* Section 16.3.6).

Examples:

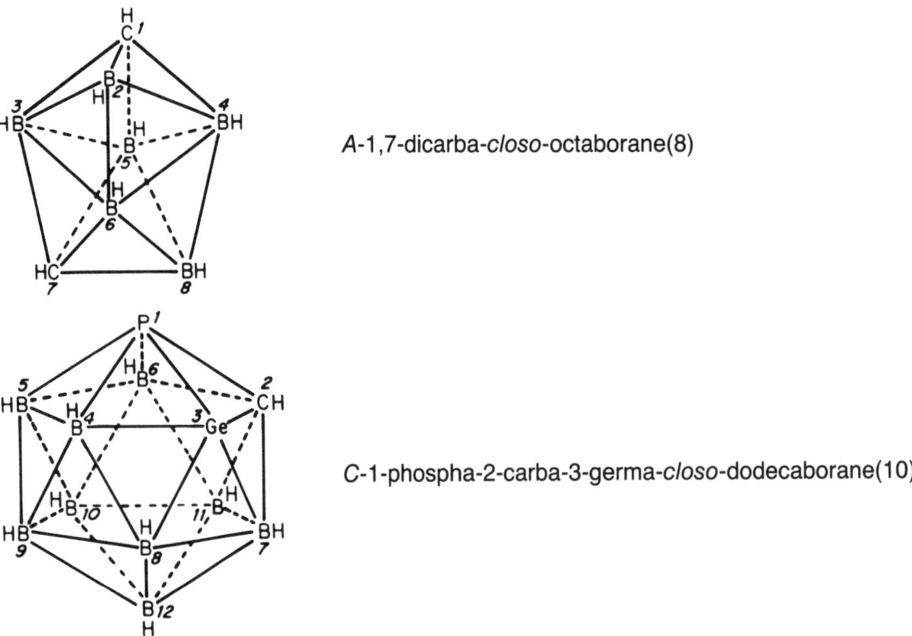


*A*-1,7-dicarba-*closo*-octaborane(8)

*C*-1-phospha-2-carba-3-germa-*closo*-dodecaborane(10)

*11.3.4.6 Nonclosed Polyhedrons.* Polyhedrons that are not fully triangulated may be described by indicating the removal of vertices from a triangulated polyhedron in either of two ways (*72b*). One way is to insert an additional symbol into the CEP descriptor for the closed polyhedral structure (*see* Section 11.3.4.1). The symbol consists of an arabic numeral equal to the number of vertices removed, preceded by the minus sign to indicate removal and followed by the italic letter *v* for vertex. This symbol is enclosed in parentheses and inserted into the descriptor immediately after the morpheme "*closo*". The locants of the positions of the vertices in the closed polyhedron (*see* Section 11.3.4.5) that are removed are denoted by superscript numbers cited after the parenthetical expression. Multiple locants are separated by commas. These locants should be as high as possible consistent with the numbering of the parent closed polyhedron. Thus, the descriptor for *nido*-undecaborane(15) (structure 4 in Chart A.1, part B) would be $(11v)\{I_h$-$(1551)$-$\Delta^{20}$-*closo*$(-1v)^{12}\}$. Indication of the number of vertices in the actual structure by the first part of the descriptor in such cases eliminates the need to calculate the number from the rest of the descriptor. The second way is to indicate the removal of a vertex from a closed polyhedral structure by attaching the subtractive prefix *n*-debor- (*n* being the locant of the missing vertex) directly in front of the name of the closed polyhedron. (The two methods of specifying the positions of bridging hydrogen atoms used in the examples that follow are both based on published proposals (*72b*) and may be used with either procedure for indicating missing vertices.)

Examples:*

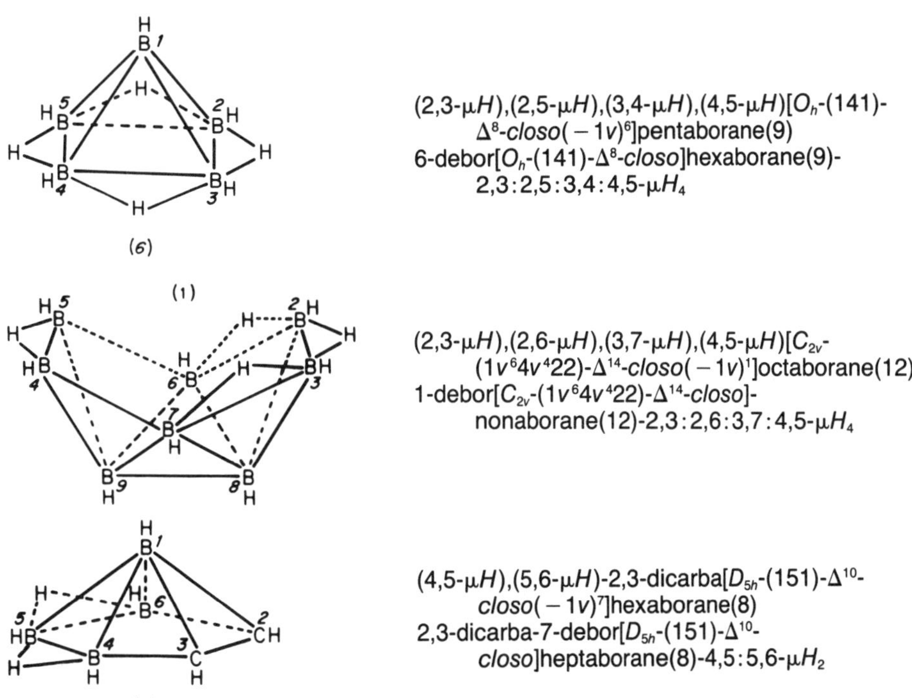


*11.3.4.7* ***Capped Polyhedrons.*** The addition of a vertex to a triangular face or an edge of a polyhedron may be indicated in the CEP descriptor by a symbolism similar to that for nonclosed structures (*see* Section 11.3.4.6). A plus sign is used to indicate addition, and sets of locants are used to describe the face or edge of the polyhedron that is capped (*72a*). Such additional vertices are numbered, beginning with the next number beyond those used to number the closed polyhedral structure (*see* Section 11.3.4.5). There are few examples of capped polyhedral polyboron hydrides, but there are a number of metallapolyboron hydrides with capped structures. The method described should also be applicable to capped nonclosed structures, but details have not yet been set forth.

*The parenthetical number represents the missing vertex.

Examples:

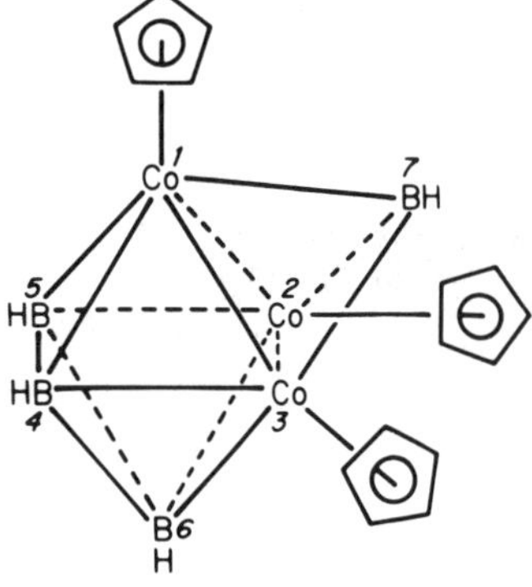

1,2,3-tris-($\eta^5$-cyclopentadienyl)-1,2,3-tricobalta[$O_h$-(141)-$\Delta^8$-*closo*-(+1*v*)$^{1,2,3}$]heptaborane(4)

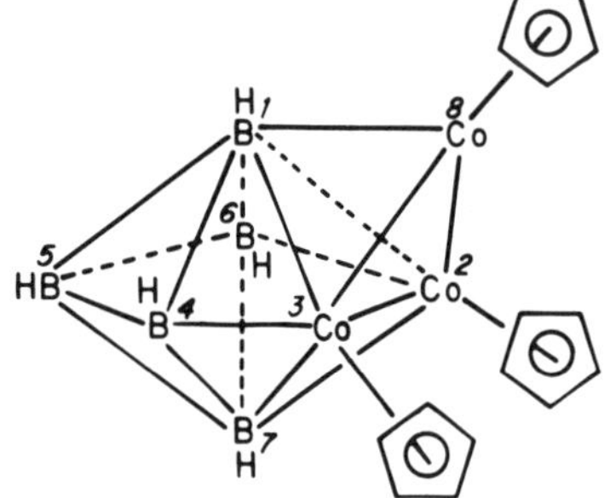

2,3,8-tris($\eta^5$-cyclopentadienyl)-2,3,8-tricobalta[$D_{5h}$-(151)-$\Delta^{10}$-*closo*-(+1*v*)$^{1,2,3}$]octaborane(5)

*11.3.4.8 Linear Conjuncto Structures.* Systems consisting of two or more closed or nonclosed polyhedrons linked in a linear fashion by direct bonding or by sharing one, two, or more atoms are called "conjuncto" (*73*) and may be described by citing consecutively the appropriate CEP descriptors for the components (*72c*). A locant description of the attachment between two components is inserted between the descriptors for those components. It consists of the locants for the polyhedral vertices (*see* Section 11.3.4.5) involved in the attachment, cited as follows:

| Type of Attachment | Designation |
|---|---|
| Two-center boron–boron bond | -a:y′- |
| One shared vertex | -a-*commo*-y′- |
| Two shared vertices (edge fusion) | -a,b-*dicommo*-x′,y′- |
| Three shared vertices | -a,b,c-*tricommo*-x′,y′,z′- |
| Bridging hydrogen | -a-μ*H*-y′-, -a,b-di-μ*H*-x′,y′-, etc. |

Examples:

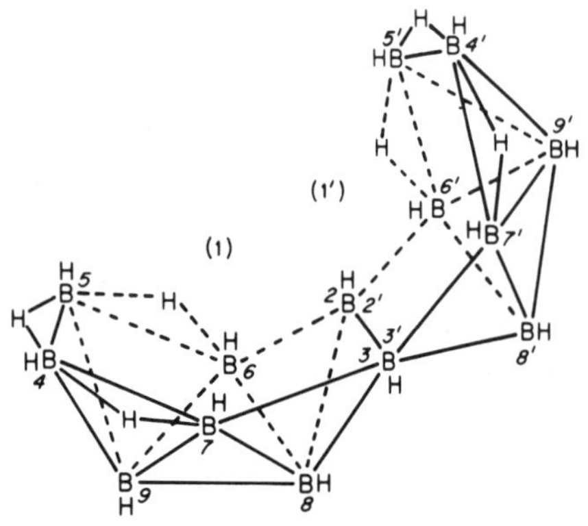

2*H*,3*H*-(4,5-μ*H*),(4′,5′-μ*H*),(4,7-μ*H*),(4′,7′-μ*H*),-(5,6-μ*H*),(5′,6′-μ*H*)[(8*v*)[$C_{2v}$-(1$v^6$4$v^4$22)-$\Delta^{14}$-*closo*(−1*v*)$^1$]-2,3-*dicommo*-2′,3′-(8*v*)[$C_{2v}$-(1$v^6$4$v^4$22)-$\Delta^{14}$-*closo*-(−1*v*)$^1$]-tetradecaborane](20)

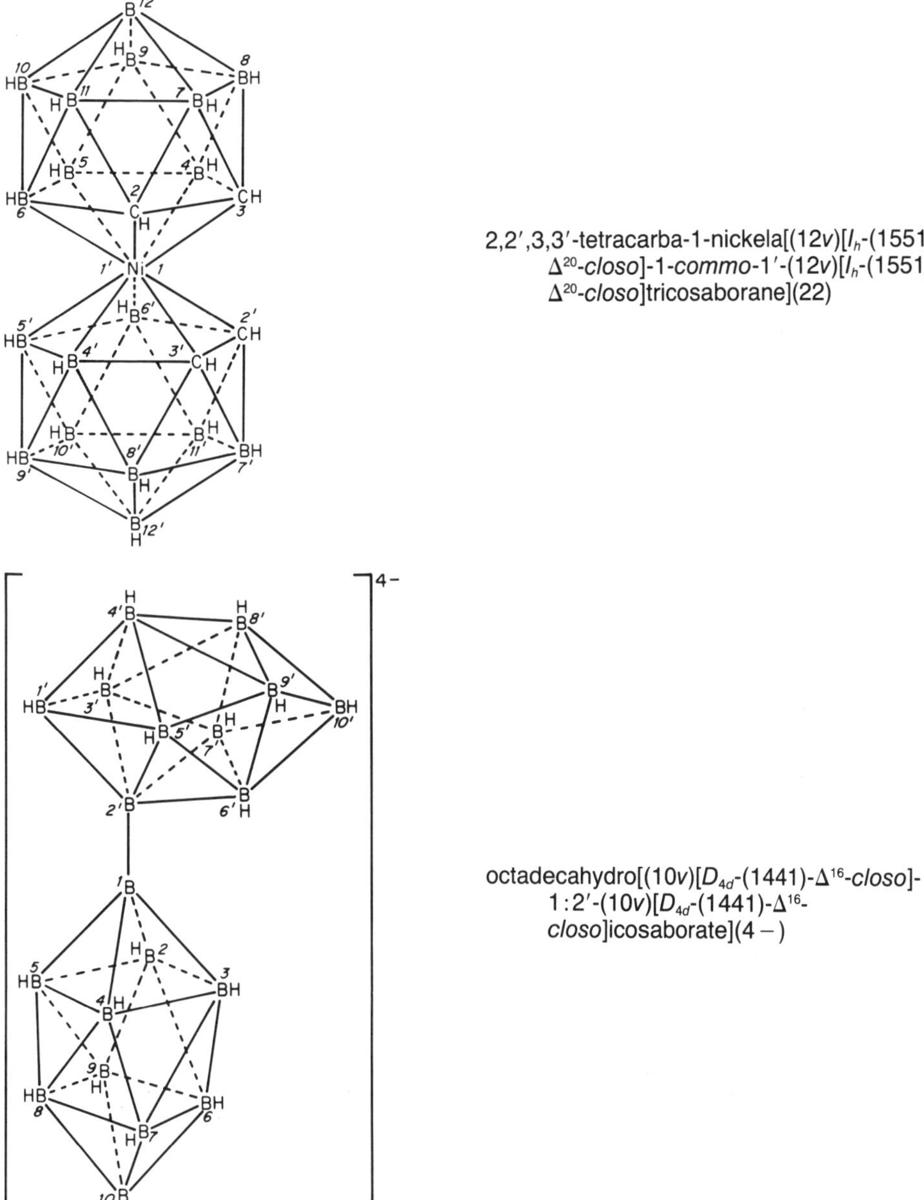

2,2′,3,3′-tetracarba-1-nickela[(12*v*)[$I_h$-(1551)-$\Delta^{20}$-*closo*]-1-*commo*-1′-(12*v*)[$I_h$-(1551)-$\Delta^{20}$-*closo*]tricosaborane](22)

octadecahydro[(10*v*)[$D_{4d}$-(1441)-$\Delta^{16}$-*closo*]-1:2′-(10*v*)[$D_{4d}$-(1441)-$\Delta^{16}$-*closo*]icosaborate](4−)

## 11.4 Derivatives of Polyboron and Heteropolyboron Hydrides

### 11.4.1 Molecular Species

Neutral polyboron and heteropolyboron hydrides in which atoms or groups have replaced hydrogen atoms of the parent hydride are named by substitutive nomenclature (*see* Chapter 9) based on the name of the parent polyboron hydride. One hydrogen atom is assumed to be bonded to each boron or carbon atom in the parent structure; the positions of other hydrogen atoms must be specified. Hydrogen atoms specified by an indicated *H* symbol (*72a*) are included in the parenthetical numerical suffix, whereas hydrogen atoms expressed by a ligand name are not. The absence of any hydrogen atom that is on a boron or carbon atom in the parent structure by implication is indicated by the prefix dehydro- and is not included in the parenthetical numerical suffix. The presence of a neutral group attached to a skeletal atom is indicated with the ligand name.

Examples:

1-chlorodiborane(6)

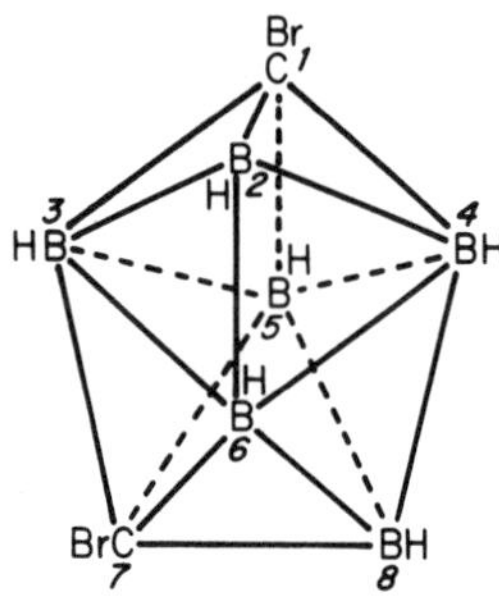

*A*-1,7-dibromo-1,7-dicarba[$D_{2d}$-(2222)-$\Delta^{12}$-*closo*]octaborane(8)

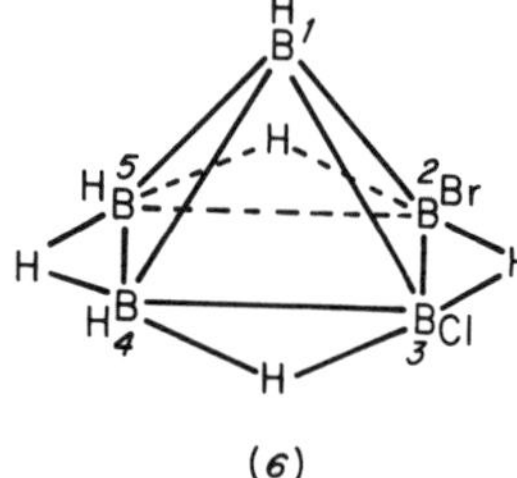

(6)

*C*-2-bromo-3-chloro-*nido*-pentaborane(9)
*C*-2-bromo-3-chloro-(2,3-$\mu H$),(2,5-$\mu H$),(3,4-$\mu H$),(4,5-$\mu H$)-[$O_h$-(141)-$\Delta^{8}$-*closo*-$(-1v)^{6}$]pentaborane(9)

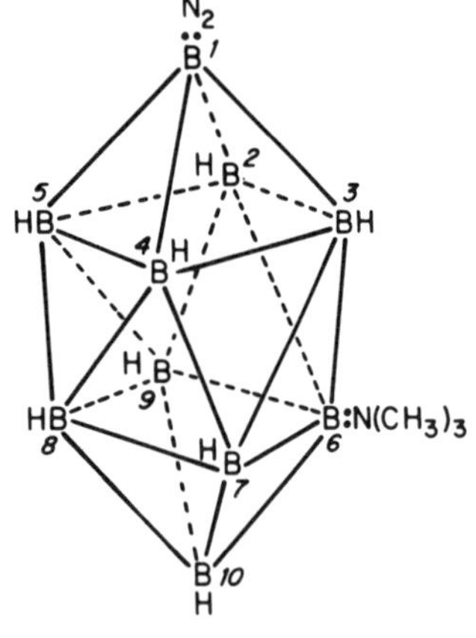

1-(dinitrogen)-6-(trimethylamine)[1,6-didehydro[$D_{4d}$-(1441)-$\Delta^{16}$-*closo*]-decaborane](8)

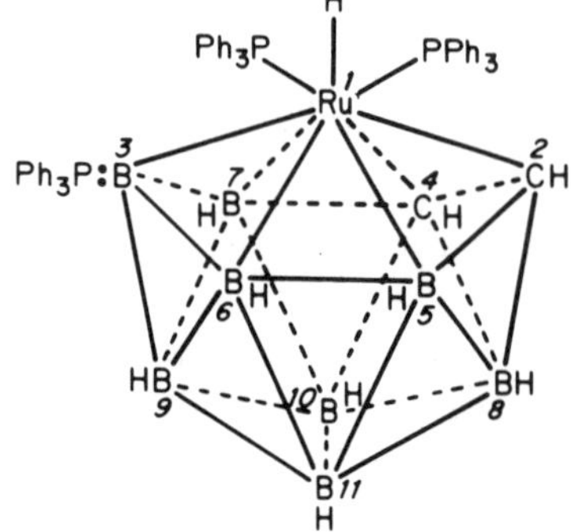

*C*-1-hydrido-1,1,3-tris(triphenylphosphine)[3-dehydro-2,4-dicarba-1-ruthena[$C_{2v}$-($1v^{6}2v^{4}422$)-$\Delta^{18}$-*closo*]undecaborane](9)

## 11.4.2 Ionic Species

Ionic polyboron and heteropolyboron hydrides and their derivatives are usually named by the principles of additive nomenclature, that is, by describing all atoms or groups, including hydrogen, as side groups (*see* Chapter 5) attached to the skeletal structure, which is described by one of the methods already given.

Examples:

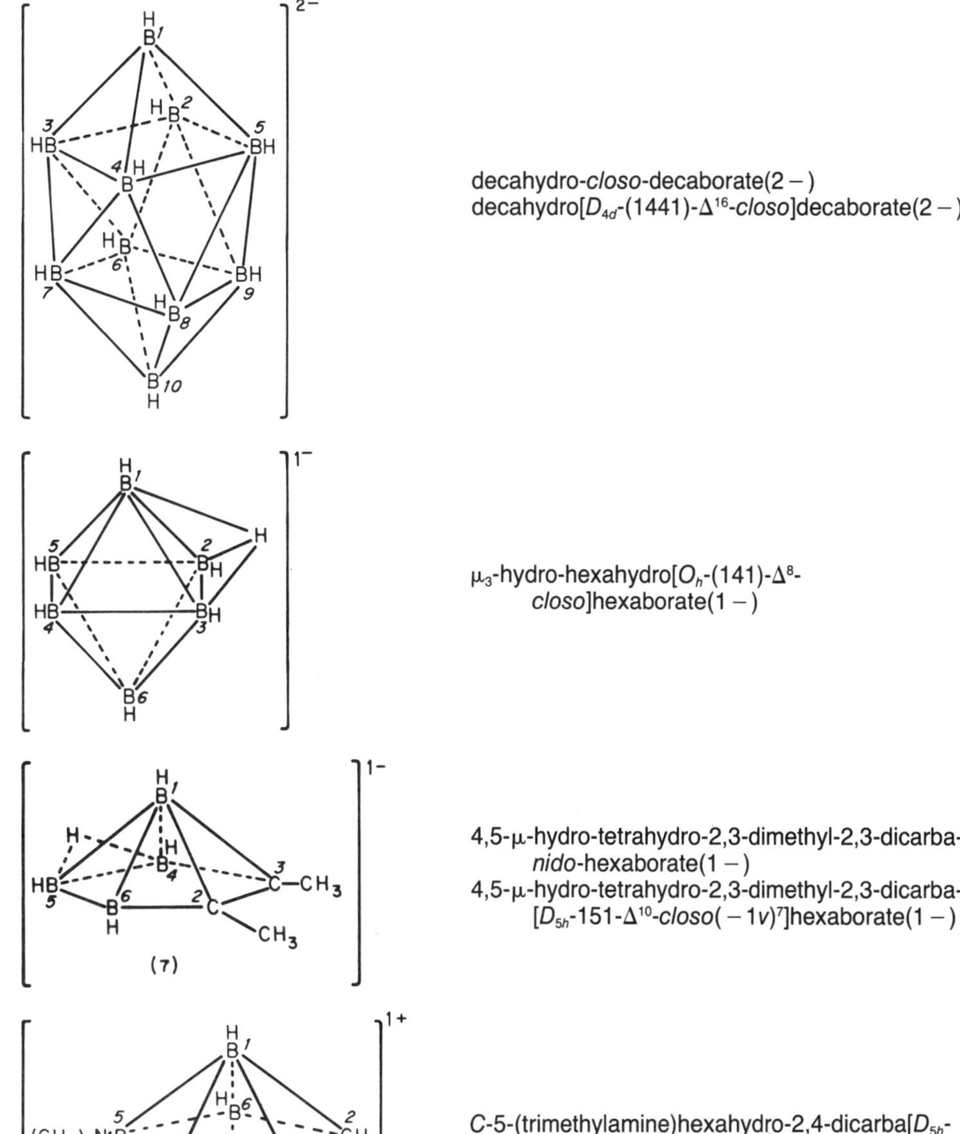


## 11.4.3 Addition Compounds

***11.4.3.1 General Discussion.*** Many neutral boron compounds may be considered, at least formally, as derived by the association of an electron pair of an atom in a neutral compound (a Lewis base) with a neutral tricoordinate boron atom in a neutral boron compound (a Lewis acid). These compounds may also be viewed as derived by the replacement of a hydride ion of an ionic boron species by a neutral atom or group. Such derivatives may be named as coordination compounds (*see* Chapter 5) or as addition compounds (*see* Chapter 13).

*11.4.3.2 Coordination-Compound Names.* Addition compounds of $BH_3$ (or $BH_4^-$) are usually readily named by additive nomenclature principles. Neutral polyboron hydrides with neutral atoms or groups attached to the boron atoms are named by additive nomenclature principles using the CEP descriptor for structural specificity or as derivatives of a dehydro polyborane, with the neutral substituents described by ligand names but with all other substituent atoms and groups expressed as in substitutive nomenclature. All groups attached to atoms of heteropolyboron hydrides other than boron or carbon are described by their names as ligands in coordination nomenclature (*see* Section 11.4.1 and examples marked "C" in Section 11.4.3.4).

*11.4.3.3 Addition-Compound Names.* Addition compounds of $BH_3$ (or $BH_4^-$) may also be given names in which the names of the constituents are separated by a dash and the boron component is cited last (*see* Section 13.2). Type-2 numerical prefixes (*see* Table A.II) are used as appropriate. Different modes of attachment may be distinguished by replacing the dash with parentheses enclosing the italicized symbols of the two atoms involved in the attachment separated by a dash. Locants of the attaching atoms may be given as superscripts to the italicized atomic symbols. If the addition is intramolecular, the parenthetical expression is placed at the end of the name of the neutral borane derivative. (*See* examples marked "A" in Section 11.4.3.4.)

*11.4.3.4 Examples:*

| | |
|---|---|
| $H_3B{\cdot}NH_3$ | amminetrihydroboron (C)<br>ammonia–borane (A) |
| $F_3B{\cdot}O(C_2H_5)_2$ | (diethyl ether)trifluoroboron (C)<br>(diethyl ether)–(boron trifluoride) (A) |
| $H_3B{\cdot}NH_2CH_2CH_2NH_2{\cdot}BH_3$ | μ-1,2-ethanediamine-κ*N*:κ*N'*-bis(trihydroboron) (C)<br>1,2-ethanediamine–bis(borane) (A) |
| $OC{\cdot}BH_3$ | (carbon monoxide-κ*C*)trihydroboron (C)<br>carbon monoxide(*C*–*B*)borane (A) |
| 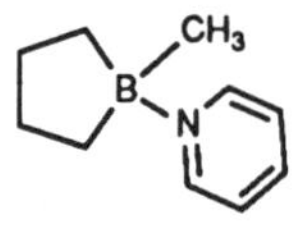 $CH_3$ B N | (1,4-butanediyl-κ$C^1$,κ$C^4$)(methyl)(pyridine)boron (C)<br>pyridine–(1-methylborolane) (A) |
| $C_6H_5NHC(O)NH_2{\cdot}BH_3$ | trihydro(1-phenylurea-κ$N^3$)boron (C)<br>1-phenylurea($N^3$–*B*)borane (A) |
| 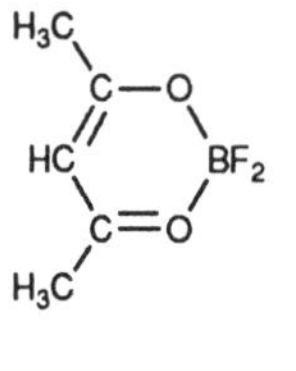 $H_3C$ C—O HC $BF_2$ C=O $H_3C$ | difluoro(2,4-pentanedionato-κ*O*,κ*O'*)boron (C)<br>difluoro[(1-methyl-3-oxo-1-butenyl)oxy]borane(*O*–*B*) (A) |

# Chapter 12
# Organometallic Compounds

## 12.1 Introduction

Organometallic compounds are characterized by the presence of at least one direct bond between a carbon atom and a metal atom. Most organometallic compounds in which there is sigma or pi or both sigma and pi bonding between the metal atom and the carbon atom(s) are readily named by the methods of additive nomenclature (*see* Sections 5.3 and 5.4). If a metal forms molecular hydrides, substitutive nomenclature may be used as an alternative procedure to name its organometallic compounds, that is, as hydrides with organic substituents (*see* Sections 9.2 and 9.3). Some of the organometallic derivatives of the very electropositive elements are saltlike and may be named as ionic compounds (*see* Section 8.2.9). Ionic nomenclature is also sometimes used for compounds in which the organometallic groups can be considered cationic. The methods of this chapter can be applied to the naming of compounds with direct bonds between carbon and all other elements.

## 12.2 Organometallic Compounds Containing Only Organic Side Groups

Names for compounds containing only organic groups and metal atoms may be formed by any of the three procedures given in Section 12.1. The hapto convention, which was outlined in Section 5.3.5.3, is used in additive nomenclature to indicate that there is linkage through two or more adjacent atoms of an organic group bonded to a metal atom. This type of bonding is indicated by the symbol $\eta^n$, in which $n$ indicates the number of atoms involved in the linkage. Bonding by the sharing of electron pairs ($\sigma$ bonds) is implied if the symbol $\eta^n$ is not used.

| Examples: | | | |
|---|---|---|---|
| | $(C_2H_5)_4Pb$ | tetraethyllead | (additive) |
| | | tetraethylplumbane | (substitutive) |
| | $(C_6H_5)_3GeGe(C_6H_5)_3$ | hexaphenyldigermane | (substitutive) |
| | $C_3H_7Li$ | propyllithium | (additive) |
| | | lithium propanide | (ionic) |
| | $C_6H_5ZnCH_3$ | methylphenylzinc | (additive) |
| | $(C_2H_5)_2Te$ | diethyltellurium | (additive) |
| | | diethyltellane | (substitutive) |
| | | diethyl telluride | (ionic) |
| | $Fe(C_5H_5)_2$ | bis($\eta^5$-cyclopentadienyl)iron | (additive) |
| | | iron(2+) cyclopentadienide | (ionic) |
| | $Cr(C_6H_6)_2$ | bis($\eta^6$-benzene)chromium | (additive) |

## 12.3 Organometallic Compounds Containing Organic and Other Side Groups

An organometallic compound in which hydrogen is bonded directly to the metal atom may be named as a substituted molecular hydride if it can be visualized conceptually as being derived from a molecular hydride (*see* Section 9.2). Other organometallic compounds of this type may be named as molecular or ionic species by additive procedures (*see* Sections 5.3, 5.4, and 8.4.1).

Examples:

| | | |
|---|---|---|
| $C_2H_5SbH_2$ | ethylstibine | (substitutive) |
| | ethyldihydridoantimony | (additive) |
| $(C_6H_5)_2SnH_2$ | diphenylstannane | (substitutive) |
| | dihydridodiphenyltin | (additive) |
| $C_2H_5BeH$ | ethylhydridoberyllium | (additive) |
| | ethylberyllium hydride | (ionic) |
| $(C_5H_5)_2ReH$ | bis($\eta^5$-cyclopentadienyl)hydridorhenium | (additive) |
| | bis($\eta^5$-cyclopentadienyl)rhenium hydride | (ionic) |
| $(C_6H_5)(C_6H_5CH_2)AuH$ | benzylhydridophenylgold | (additive) |
| | benzyl(phenyl)gold hydride | (ionic) |

Organometallic compounds that contain substituents other than organic groups or hydrogen bonded directly to the metal atom may also be named by the preceding methods, as appropriate to the type of compound.

Examples:

| | | |
|---|---|---|
| $CH_3MgI$ | iodo(methyl)magnesium | (additive) |
| | methylmagnesium iodide | (ionic) |
| $(CH_3)_3SnCl$ | chlorotrimethylstannane | (substitutive) |
| | chlorotrimethyltin | (additive) |
| $(C_6H_5)_3SbCl_2$ | dichlorotriphenyl-$\lambda^5$-stibane | (substitutive) |
| | dichlorotriphenylantimony | (additive) |
| | triphenylantimony(2+) chloride | (ionic) |
| $(C_5H_5)_2TiCl_2$ | dichlorobis($\eta^5$-cyclopentadienyl)titanium | (additive) |
| | bis($\eta^5$-cyclopentadienyl)titanium(2+) chloride | (ionic) |
| $CH_3SnH_2Cl$ | chloro(methyl)stannane | (substitutive) |
| | chlorodihydridomethyltin | (additive) |
| | methyltin(3+) chloride dihydride | (ionic) |
| $C_6H_5HgOH$ | hydroxo(phenyl)mercury | (additive) |
| | phenylmercury hydroxide | (ionic) |

Organometallic coordination compounds may be named by the usual additive procedures (*see* Section 5.3).

Examples:

| | |
|---|---|
| $[Pt(CH_3)Cl\{P(C_2H_5)_3\}_2]$ | chloro(methyl)bis(triethylphosphine)platinum |
| $[Sb(CF_3)_3(C_5H_5N)]$ | (pyridine)tris(trifluoromethyl)antimony |
| $[Fe(C_6Cl_5)_2\{P(C_6H_5)_3\}_2]$ | bis(pentachlorophenyl)bis(triphenylphosphine)iron |
| $[Ru\{(CH_3)_2PC_2H_4P(CH_3)_2\}_2H(C_6H_5)]$ | bis[1,2-ethanediylbis(dimethylphosphine)]-hydrido(phenyl)ruthenium [A replacement name, bis(2,5-dimethyl-2,5-diphosphahexane)hydrido(phenyl)ruthenium, may be encountered.] |
| $[(C_2H_5)_2AlCl]_2$ | di-$\mu$-chloro-tetraethyldialuminum |

## 12.4 Side Groups Derived from Organometallic Compounds

Side groups derived from organometallic compounds may be named in one of two ways when the side group forms bonds through the metal atom. If the side group may be formally derived by the loss of hydrogen from a molecular hydride, the procedures of Section 9.3 may be used.

Examples:

| | |
|---|---|
| $(C_6H_5)_3Sn–$ | triphenylstannyl |
| $(CH_3)_2Ge=$ | dimethylgermylidene<br>dimethylgermanediyl |

If the metal atom does not form a molecular hydride, names are formed by additive procedures after adding the suffix -io to a stem name for the metal atom (*see* Table A.I).

Examples:

| | |
|---|---|
| $CH_3Hg–$ | methylmercurio |
| $(C_6H_5)_2Tl–$ | diphenylthallio |

This type of nomenclature may also be used to indicate the presence of a metal as a monovalent substituent in an organometallic compound, rather than as the central atom of an organometallic compound.

Examples:

| | |
|---|---|
| $(C_6H_5)_3CNa$ | triphenyl(sodio)methane<br>[an alternative to (triphenylmethyl)sodium] |
| $C_6H_5C{\equiv}CCu$ | cuprio(phenyl)ethyne<br>[an alternative to (phenylethynyl)copper(I)] |
| $LiCH_2C{\equiv}CLi$ | 1,3-dilithiopropyne |

Organometallic side groups that function as bridging groups between central atoms may be so indicated in the usual manner (*see* Section 6.2).

Example:

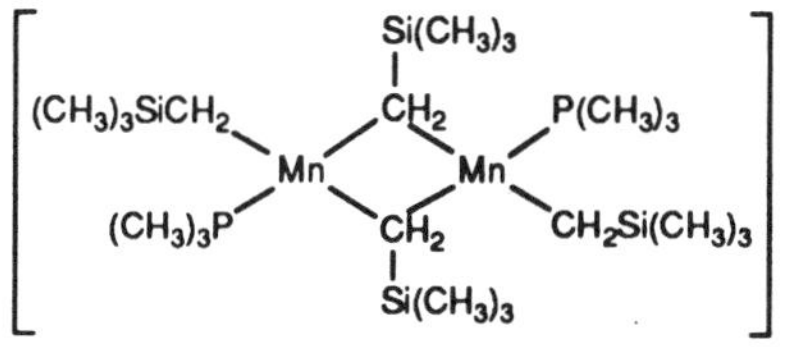

$[(CH_3)_3SiCH_2)(CH_3)_3P)Mn(\mu\text{-}CH_2Si(CH_3)_3)_2Mn(P(CH_3)_3)(CH_2Si(CH_3)_3)]$

bis-[μ-(trimethylsilyl)methyl]-bis{(trimethylphosphine)-[(trimethylsilyl)methyl]manganese(II)}

## 12.5 Ions Derived from Organometallic Compounds

An organometallic cation that is formally derived from a parent cationic hydride (*see* Section 9.4) may be named as a substituted derivative of the parent cation. Likewise, an organometallic anion that is formally derived from a parent anionic hydride (*see* Section 9.4) may be named as a substituted derivative of the parent anion.

Examples:

| | |
|---|---|
| $Bi(C_2H_5)_4^+$ | tetraethylbismuthonium |
| $Sn(C_6H_5)_3^+$ | triphenylstannylium<br>triphenyl-λ²-stannanium |
| $Ge(CH_3)_3^-$ | trimethylgermanide |

Organometallic cations based on metal atoms that do not form molecular hydrides and organometallic anions that are formally derived by the addition of one or more hydride ions to a molecular hydride or by the addition of anionic ligands to a central atom may be named by additive procedures (*see* Section 5.3).

Examples:

| | |
|---|---|
| $(CH_3)_2Tl^+$ | dimethylthallium(1+) ion |
| $(C_2H_5)_3Pt^+$ | triethylplatinum(1+) ion |
| $B(C_6H_5)_4^-$ | tetraphenylborate(1−) ion |
| $AlH(CH_3)_3^-$ | hydridotrimethylaluminate(1−) ion |
| $Sb(C_6H_5)Cl_5^-$ | pentachloro(phenyl)antimonate(1−) ion |

Substitutive names may be used for anions of oxo acids (and also for the acids) in which organic groups are present in place of hydrogen atoms bonded directly to the central atom (*see* Section 8.2.3).

Examples:

| | |
|---|---|
| $C_2H_5As(O)O_2^{2-}$ | ethylarsonate(2−) ion |
| $(CH_3)_2PO^-$ | dimethylphosphinite(1−) ion |

## Chapter 13

# Addition Compounds

## 13.1 Introduction

The reaction of two molecular species to form a third species is a common phenomenon that leads to a variety of combinations. Many of the products are given names in accord with the resulting structure. However, a number of products (loosely called "addition compounds"), either for convenience or because of difficulty in specifying the structure, are named in accord with the stoichiometry of the union. This general category includes solvates, molecular or electron pair donor–acceptor compounds, clathrates or inclusion compounds, and intercalate or laminar compounds. Various general terms have been applied to addition compounds, the most common being "complex" and "adduct". The first term causes confusion with coordination compounds, the latter with products in reactions such as the Diels–Alder reaction.

The concept of addition compounds goes back to the dualistic conceptions of Lavoisier and Berzelius about the nature of chemical combinations leading to acids, bases, and salts. They formulated a salt such as $Na_2SO_4$, for example, as $Na_2O{\cdot}SO_3$ and its hydrate $Na_2SO_4{\cdot}10H_2O$ as $Na_2O{\cdot}SO_3{\cdot}10H_2O$. Mineralogists, ceramists, and metallurgists still use such formulations, but chemists confine such formulas to addition compounds, especially solvates. In general terms, the formula for an addition compound is written $nA{\cdot}pB{\cdot}qC\ldots$, in which $n$ specifies the number of units of A present in the empirical formula, $p$ the number of units of B, $q$ the number of units of C, etc. It is easy to overlook the fact that a coefficient applies to everything preceding the centered point in complex formulas such as $2\{[(CH_3)_2CH]_2N\}_3P{\cdot}AlCl_3$, so the alternative formula $\{\{[(CH_3)_2CH]_2N\}_3P\}_2{\cdot}AlCl_3$ is sometimes used.

## 13.2 General Practice

Names of addition compounds may be formed by listing the names of all the constituents separated one from another by a dash (*see* Section 2.2.6.4), formerly a spaced hyphen (*22s*). The number of each constituent is indicated by arabic numerals separated by slashes (*see* Section 2.2.6.7) and enclosed in parentheses attached to the end of the full name. CAS still follows the practice of using a colon as a separator, a use not recognized in the 1970 IUPAC Inorganic Rules. The use of the slash or solidus as a separator has been criticized because the expression could be interpreted as a fraction. Alternatively, the ratio of constituents may be indicated by appropriate numerical prefixes (*see* Table A.II). Although there is no general agreement on the order of citing the constituents of an addition compound, the 1970 IUPAC Inorganic Rules give this general instruction

(*22s*): "Boron compounds and water are always cited last in the order. Other molecules are cited in order of increasing number; any which occur in equal numbers are cited in alphabetical order." Another procedure sometimes used is to list the donor molecule(s) first.

Examples:

| | |
|---|---|
| $BiCl_3 \cdot 3PCl_5$ | bismuth trichloride–phosphorus pentachloride(1/3) |
| $BF_3 \cdot 2H_2O$ | boron trifluoride–water(1/2) |
| $Al_2(SO_4)_3 \cdot K_2SO_4 \cdot 24H_2O$ | aluminum sulfate–potassium sulfate–water(1/1/24) |
| $NH_3 \cdot B_3H_7$ | ammonia–triborane(7)(1/1) |
| $2(CH_3)_2S \cdot B_{10}Cl_6H_8$ | bis(dimethyl sulfide)–hexachlorodecaborane(14) |

If there are possible alternatives in the manner of attachment between two molecules or if it is desired to emphasize the atoms involved in the sharing of a pair of electrons, the dash is replaced by parentheses enclosing the italicized symbols of the two elements forming the bond with a dash between the symbols.

Examples:

| | |
|---|---|
| $OC \cdot BH_3$ | carbon monoxide(*C–B*)borane(1/1) |
| $CH_3ONH_2 \cdot BH_3$ | *O*-methylhydroxylamine(*N–B*)borane(1/1) |

This general system for naming addition compounds is not used by CAS. Instead the addition compound is named by citing one constituent followed by "compound with", the name(s) of the other constituent(s), and a parenthetical expression containing the relative number of constituents separated by colons as necessary. Molecular addition compounds of neutral components are generally indexed at the name of each component, with the exception of certain common components specified in the CA *Index Guide* (*32c*).

Examples:

| | |
|---|---|
| $I_2 \cdot POCl_3$ | phosphoryl chloride compd. with iodine(1:1)<br>iodine compd. with phosphoryl chloride(1:1) |
| $AlCl_3 \cdot 4C_2H_5OH$ | aluminum chloride ($AlCl_3$) compd. with ethanol(1:4)<br>ethanol compd. with aluminum chloride($AlCl_3$)(4:1) |

## 13.3 Solvates

Many salts crystallize from a solvent to yield crystals containing molecules of the solvent as well as the salt. These crystalline products are known as solvates, and the solvent component is known as solvent of crystallization. The terms hydrate and ammoniate have been widely used for solvates containing water and ammonia, respectively. Because the suffix -ate is now a general suffix for anions (*see* Section 8.4.1), it should not be used in naming solvates. For example, alcoholates are the salts of alcohols, so ethanolate should not be used for products containing ethanol of crystallization. CNIC recognizes that the use of "hydrate" has to be accepted as an exception to the restriction of the suffix -ate to anions, in view of the widespread use of that term for compounds containing water of crystallization (*22s*). It does, however, restrict the use of hydrate to the designation of water bound in an unspecified way and recommends that specific additive names be used whenever the structure of the hydrate is known. Both hydrate and ammoniate are used in the CA indexes.

Examples:

| | |
|---|---|
| $Na_2CO_3 \cdot 10H_2O$ | sodium carbonate decahydrate<br>sodium carbonate–water(1/10) |

| | |
|---|---|
| $[Cu(H_2O)_4]SO_4 \cdot H_2O$ | tetraaquacopper(II) sulfate monohydrate [preferred to copper(II) sulfate pentahydrate] |
| $3CdSO_4 \cdot 8H_2O$ | cadmium sulfate–water(3/8) |
| $CdCl_2 \cdot 8NH_3$ | cadmium chloride–ammonia(1/8) |
| $BF_3 \cdot 2H_2O$ | boron trifluoride dihydrate<br>boron trifluoride–water(1/2) |

## 13.4 Clathrates

Some substances have the ability to crystallize either from the liquid phase or from solution in such a way as to leave either cavities or channels that may enclose another substance. Such substances that contain other substances have been called "enclosure compounds" or "clathrates". The continuous lattice is often called the "host" and the enclosed substance the "guest". The general procedure for naming addition compounds (*see* Section 13.2) is used for naming clathrates.

Examples:

| | |
|---|---|
| $Kr \cdot 6H_2O$ | krypton–water(1/6) |
| $COS \cdot 2H_2S \cdot 17H_2O$ | carbonyl sulfide–hydrogen sulfide–water(1/2/17) |
| $C_6H_6 \cdot NH_3 \cdot Ni(CN)_2$ | ammonia–benzene–nickel(II) cyanide(1/1/1) |
| $6Br_2 \cdot 46H_2O$ | bromine–water(6/46) |

The nomenclature of zeolites (the so-called "molecular sieves"), which have rigid frameworks with wide windows between relatively large cavities, has been discussed elsewhere (*74*).

## 13.5 Intercalation Compounds

It is possible for some substances to insert themselves between the layers of crystalline substances to yield stable materials with a wide range of compositions. There is usually an upper limit to the amount of material that can be held between layers, but this is not always realized in a given instance. The definite formulas that can be written for the different situations are termed "limiting formula" and "average formula". Occasionally definite stoichiometric compounds of intermediate composition are formed. There is currently no way to indicate that a material is an intercalated species, so either stoichiometric names, when appropriate, or general additive names (*see* Section 13.2) are used for intercalation compounds.

Examples:

| | |
|---|---|
| $(CO)_x$ | graphitic oxide |
| $(CF)_x$ | carbon monofluoride |
| $C_8K$ | potassium–graphite(1/8) |
| $TaS_2 \cdot NH_3$ | tantalum disulfide–ammonia(1/1) |
| $Na_xTaS_2$ ($x$ = 0.4–0.7) | sodium–tantalum disulfide(0.4–0.7/1) |

# Chapter 14
# Nonstoichiometric Species

## 14.1 Introduction

The preceding chapters have been concerned for the most part with the naming of molecular and ionic species with small whole-number atomic ratios. This chapter will be devoted to nonstoichiometric species. In some cases the composition of substances deviates from simple stoichiometry, and in others it is variable. Substances with variable composition were called "berthollides" in the older literature (*22t*) to distinguish them from substances with a fixed simple composition, which were called "daltonides". These terms are not encountered very often now, perhaps because a variety of causes for nonstoichiometry have become apparent. At present there are no universally accepted procedures for dealing with the various situations encountered, and detailed nomenclature procedures still need to be developed.

All solids except amorphous and glassy ones have an orderly arrangement of constituent atoms or ions known as a lattice. Normally neither the name nor the formula specifies the nature of the lattice or distinguishes among solids composed of molecular units, ionic units, or single giant molecules. Infinite lattices may be indicated by writing formulas such as $(NaCl)_\infty$ and $(SiO_2)_\infty$. On a micro scale, a lattice may exhibit a number of alterations leading to solid-state species that do not exhibit simple atomic ratios. Consequently, these species require nomenclature systems quite different from systems pertaining to stoichiometric substances. Some of these alterations will be discussed next.

There may be regions of disorder in a lattice, or there may be boundaries between ordered regions. Lattice sites may be occupied in part by foreign atoms or ions without significant distortion of the lattice, resulting in a solid solution. A solid solution may form over all concentrations (infinitely variable) or over a limited range of concentrations. The substitution into a lattice of a foreign ion for a lattice ion may result in vacant lattice sites when the ions involved do not have the same charge. Within the lattices of many metals and compounds there are small unoccupied regions known as interstitial sites or holes. Such sites (which are usually designated as tetrahedral, octahedral, or dodecahedral, depending upon the symmetry of the atoms defining the vacant sites) may also be occupied by foreign atoms, ions, or molecules. A pair of metals may yield a series of phases, each of which has an ideal formula, but with considerable variation on both sides of the ideal composition within each phase.

A distinctive nomenclature for each of the preceding situations, although desirable, currently does not exist. In practice the nomenclature used depends to a large extent on the amount of information the name is intended to convey. In many cases, formulas serve better than names to describe the situation.

## 14.2 Qualitative Indication of Nonstoichiometry

### 14.2.1 Formulas

The sign ≈ (read as "circa") may be placed in front of a formula to indicate that the composition of the species varies over an appreciable range and is not precisely the composition called for by the formula. The circa sign has also been printed above the formula, but that method of notation seems to be losing favor.

Examples: $\approx FeS$ $\overset{\approx}{CuZn}$

When the variable composition within a phase is caused wholly or partially by replacement, symbols for the atoms or atomic groups and for the atoms or groups that replace them are separated by a comma and placed together inside a set of parentheses. The resulting expression is then inserted into the formula for the parent species in place of the appropriate atomic symbol(s). If it is possible, the formula should be written so that the limits of the homogeneity range are represented when one or the other of the two atoms or groups is eliminated from the parenthetical expression. If only part of the homogeneity range is referred to, the major constituent should be placed first.

Examples:

| | |
|---|---|
| (Cu,Ni) | complete range between Cu and Ni |
| K(Br,Cl) | complete range between KBr and KCl |
| $(Li_2,Mg)Cl_2$ | complete range from $Li_2Cl_2$ (or LiCl) to $MgCl_2$ |
| $Al_6(Al_2,Mg_3)O_{12}$ | complete range from $Al_8O_{12}$ (or $Al_2O_3$) to $Al_6Mg_3O_{12}$ (or $Al_2MgO_4$) |

The system formed between copper and zinc exhibits phases that have the ideal compositions CuZn, $Cu_5Zn_8$, and $CuZn_3$, but the individual phases are stable on both sides of these ideal compositions. The intermediate ideal compositions have the unique formulas given and thus pose no problem for formulation or nomenclature. The composition range between the ideal compositions Cu and CuZn may be indicated by the formula Cu(Cu,Zn), that between CuZn and $Cu_5Zn_8$ by $Cu_{10}(Cu,Zn)_3Zn_{13}$, that between $Cu_5Zn_8$ and $CuZn_3$ by $Cu_{13}(Cu,Zn)_7Zn_{32}$, and that between $CuZn_3$ and Zn by $(Cu,Zn)Zn_3$. The total number of atoms in the formula for each variable phase is obtained by multiplying the total number of atoms in the formulas for the two limiting phases by each other. The number of variable atoms is then determined by examining the difference between the two limiting phases expressed in terms of this number.

For example, the formula for the variable phase between CuZn (2 atoms) and $Cu_5Zn_8$ (13 atoms) will contain 2 × 13 or 26 atoms varying from $Cu_{13}Zn_{13}$ to $Cu_{10}Zn_{16}$. The transition from $Cu_{13}Zn_{13}$ to $Cu_{10}Zn_{16}$ involves the gradual replacement of three Cu atoms by three Zn atoms, so the formula for the variable phase is $Cu_{10}(Cu,Zn)_3Zn_{13}$.

### 14.2.2 Names

Although strictly logical names for nonstoichiometric species can become quite cumbersome and are generally to be avoided, the procedure illustrated by the following examples is convenient when only a qualitative indication of nonstoichiometry is desired.

Examples: iron(II) sulfide (iron deficient)
molybdenum dicarbide (excess of carbon)

## 14.3 Indication of Extent of Replacement

The variables that define the extent of replacement or deficiency of atoms or groups of atoms in a given species may be indicated by inserting appropriate variables into the formula showing all atoms involved in the species in question.

Examples:

| | |
|---|---|
| $Cu_xNi_{1-x}$ | equivalent to (Cu,Ni); $x$ cannot exceed 1 |
| $KBr_xCl_{1-x}$ | equivalent to K(Br,Cl); $x$ cannot exceed 1 |
| $Li_{2-2x}Mg_xCl_2$ | equivalent to $(Li_2,Mg)Cl_2$, but shows explicitly that one vacant cation position appears for every replacement of two $Li^+$ with one $Mg^{2+}$; $x$ cannot exceed 1 |
| $Al_6Al_{2(1-x)}Mg_{3x}O_{12}$ | shows that this phase between $Al_2MgO_4$ and $Al_2O_3$ cannot contain more Mg than that corresponding to $Al_2MgO_4$ because $x$ cannot exceed 1 |
| $Fe_{1-x}O$ | indicates vacant cation sites |
| $Na_xWO_3$ | indicates interstitial $Na^+$ |

To show that the variable denoted by $x$ can attain only small values, $\delta$ or $\epsilon$ may be substituted for $x$. A particular composition may be indicated by citing the actual value of $x$, $\delta$, or $\epsilon$, which may be placed in parentheses after the general formula. The value of the variable may also be introduced into the formula itself. This notation may be used for either substitutional or interstitial solid solutions.

Examples: $Fe_{3x}Li_{4-x}Ti_{2(1-x)}O_6$ ($x = 0.35$) or $Fe_{1.05}Li_{3.65}Ti_{1.30}O_6$

Comment: The first formula may be preferable to the second because it makes the relationships easier to understand.

| | |
|---|---|
| $PdH_x$ ($x < 0.1$) | solution of hydrogen in palladium |
| $PdH_x$ ($0.5 < x < 0.7$) | palladium hydride phase |
| $Fe_{1-\delta}S$ | |
| $Al_4Th_8H_{15.4}$ | |
| $LaNi_5H_x$ ($0 < x < 6.7$) | |

## 14.4 Irregularity in Crystallographic Sites

### 14.4.1 General Discussion

As the knowledge about nonstoichiometric species increases, the phenomena leading to the nonstoichiometry become more fully understood. Defects in the crystal lattice arising from the occurrence of vacant sites or the occupation of interstitial sites are frequently involved. Additional symbolism is necessary to indicate the presence of such defects.

### 14.4.2 Designation of Sites

The symbolism used for the designation of vacant and interstitial sites is still evolving. The older recommended system is based on the use of the symbol □ to represent a site in the ideal lattice and the symbol △ to represent an interstitial site. These symbols may be modified to show that a site is cationic (□cat), anionic (□an), tetrahedral (□tet), octahedral (□oct), or cubic (□cub). Sometimes arbitrary designations, such as □a and □b for tetrahedral and octahedral sites, have been used. Additional specificity is possible by citing the point group symbol showing the symmetry of the immediate environment of the site and its coordination number in brackets after the site symbol; for example, $\square[O_h;6]$. An atom A in the site □ is indicated by the symbol $(A|\square)$, and a vacant site is indicated by the unmodified symbol □. The distribution of $n$ atoms of A over $m$ □ sites is indicated by the symbol $(A_n|\square_m)$, in which $m$ is necessarily larger than $n$. This implies that $(m - n)$ sites are vacant, and it is not necessary to show them specifically. The use of the symbol □surf or △surf to designate a site in the surface has also been suggested as useful in the discussion of catalytic reactions.

The complexity of the formulas resulting from the use of the symbols defined in the preceding paragraph has worked against their general acceptance. An alternative symbolism has been developed, based on the assumption that the atomic symbols in a parent formula indicate not only the presence of given atoms but also their location in space. The atomic symbol for each atom in the parent formula is used as a subscript to indicate the site being occupied by a given atom, and the number of atoms in the site is indicated in the same subscript by the appropriate number separated from the atomic symbol by a comma. Thus, an atom A on a site normally occupied by A in the parent structure would be indicated by $A_A$, whereas an atom A on a site normally occupied by B would be indicated by $A_B$. An atom A occupying an interstitial site is indicated by using the subscript i with the atomic symbol; that is, $A_i$ signifies that an atom A is on an interstitial site. A vacant site is indicated by the symbol $V$ with an appropriate subscript to identify its location in the lattice; that is, $V_A$ symbolizes a vacancy at a site normally occupied by A.

Examples:

LiCl with Mg and vacancies on Li sites

$$Li_{Li,1-2x}Mg_{Li,x}V_{Li,x}Cl$$
$$Li_{1-2x}Mg_x\square_xCl$$
$$(Li_{1-2x},Mg_x|\square)Cl$$

$Al_2MgO_4$ with Al and vacancies on Mg sites

$$Al_{Al,6}Al_{Mg,2x}Mg_{Mg,3-3x}V_{Mg,x}O_{12}$$
$$(Al|\square oct)_6(Al_{2x},Mg_{3-3x}|\square tet_3)O_{12}$$

Compositions intermediate between magnetite, $Fe_3O_4$, and maghemite, $\gamma$-$Fe_2O_3$

$$Fe^{II}{}_{Fe^{II,III},1-x}Fe^{III}{}_{Fe^{II,III},1+2/3x}Fe^{III}{}_{Fe^{III},1}V_{Fe^{II,III},1/3x}O_4$$
$$(Fe^{II}{}_{1-x},Fe^{III}{}_{1+2/3x}|\square oct_2)(Fe^{III}|\square tet)O_4$$

$Mg_2Sn$ with some Mg on Sn sites and vice versa

$$Mg_{Mg,2-x}Mg_{Sn,y}Sn_{Mg,x}Sn_{Sn,1-y}$$
$$(Mg_{2-x},Sn_x|\square tet_2)(Mg_ySn_{1-y}|\square cub)$$

$CaF_2$ with some F on interstitial sites instead of F sites

$$Ca_{Ca,1}F_{F,2-x}F_{i,x}V_{F,x}$$
$$Ca(F_{2-x}|\square_2)(F_x|\triangle)$$

AgBr with some Ag on interstitial sites instead of Ag sites (Frenkel defects)

$$Ag_{Ag,1-\delta}Ag_{i,\delta}V_{Ag,\delta}Br$$
$$Ag_{1-\delta}\square_\delta(Ag_\delta|\triangle)Br$$

KCl with cation and anion vacancies (Schottky defects)

$$K_{K,1-\delta}V_{K,\delta}Cl_{Cl,1-\delta}V_{Cl,\delta}$$
$$(K_{1-\delta}\square_\delta)(Cl_{1-\delta}\square_\delta)$$

### 14.4.3 Designation of Charge

When the total charge associated with an atomic position (the absolute charge) needs to be specified, the usual conventions are used (*see* Sections 4.4.2 and 4.4.3). Charges associated with nonstoichiometric species are preferably defined with respect to the ideal species and are called "effective charges". One unit of positive effective charge is indicated by a right superscript dot, $^{\bullet}$, one unit of negative effective charge by a right prime, $'$, and $n$ units of charge by $^{n\bullet}$ or $^{n\prime}$. Two units of effective charge may be indicated by doubling the charge symbol (i.e., $^{\bullet\bullet}$ or $''$, or by using $^{2\bullet}$ or $^{2\prime}$). Sites that have no effective charge relative to the unperturbed lattice may be indicated explicitly by a superscript cross, $^{\times}$. The sum of the absolute charges and effective charges must be zero for a neutral species.

Examples:

$\mathrm{Li_{Li,1-2\mathit{x}}Mg^{\bullet}_{Li,\mathit{x}}}V'_{\mathrm{Li},x}\mathrm{Cl_{Cl}}$

Comments: The Li sites occupied by Mg have an effective charge of 1+ and the vacant Li sites an effective charge of 1−. $\mathrm{Li^{\times}_{Li,1-2\mathit{x}}Mg^{\bullet}_{Li,\mathit{x}}}V'_{\mathrm{Li},x}\mathrm{Cl^{\times}_{Cl}}$ has the same meaning.

$\mathrm{Y_{Y,2-\mathit{x}}Zr^{\bullet}_{Y,\mathit{x}}O_{O,3}O''_{i,0.5\mathit{x}}}$

Comments: Every Y site occupied by Zr has an effective charge of 1+. For every two Y sites occupied by Zr, there is an interstitial O with an effective charge of 2−.

$\mathrm{Ag_{Ag,1-\delta}Ag^{\bullet}_{i,\delta}}V'_{\mathrm{Ag},\delta}\mathrm{Br_{Br}}$

Comment: For every interstitial Ag with an effective charge of 1+, there is a vacant Ag site with an effective charge of 1−.

To emphasize the semiconductor properties of a substance, it is necessary to be able to specify the presence of free electrons and the concomitant positive holes required by bulk neutrality. The symbols $\mathrm{e}'$ and $\mathrm{h}^{\bullet}$ may be used, respectively, to indicate their presence. Other symbols in use are $\mathrm{e}^-$ (or n) instead of $\mathrm{e}'$ and $\mathrm{v}^-$ (or p) instead of $\mathrm{h}^{\bullet}$. The use of n and p is not recommended because they are commonly used to designate the neutron and proton, respectively.

Examples:

| | |
|---|---|
| $\mathrm{Ge_{Ge,1-\delta}As^{\bullet}_{Ge,\delta}e'_{i,\delta}}$ | (Ge doped with As) |
| $\mathrm{Ge_{Ge,1-\delta}Ga'_{Ge,\delta}h^{\bullet}_{i,\delta}}$ | (Ge doped with Ga) |
| $\mathrm{Zn_{Zn}Zn^{2\bullet}_{i,\delta}O_{O}e'_{i,2\delta}}$ | (ZnO doped with Zn) |
| $\mathrm{Na_{Na}Cl_{Cl,1-\delta}}V^{\bullet}_{\mathrm{Cl},\delta}\mathrm{e'_{Cl,\delta}}$ | (electron trapped in vacant Cl site) |

## 14.5 Nonstoichiometric Phases

Improvement in the precision with which structures have been determined has led to the discovery of nonstoichiometric phases referred to as homologous series, noncommensurate and semicommensurate structures, vernier structures, crystallographic shear phases, Wadsley defects, chemical twinned phases, infinitely adaptive phases, and modulated structures. Phases that fall into these classes but that have essentially fixed compositions are regarded as stoichiometric, even though their formulas (such as $W_{17}O_{47}$) do not exhibit small whole-number atomic ratios. Procedures for designating the presence of these various structures have not proceeded much beyond the recognition state, so their nomenclature will not be pursued here.

Nonstoichiometric phases may also be amorphous, with all the variations inherent to that state. The appropriate symbolism (*see* Section 3.3.3) may be used in conjunction with any compositional formulas if so desired.

# Chapter 15
# Isotopically Modified Species

## 15.1 Introduction

Each element in a given chemical species is assumed to have its natural isotopic composition (*75*) unless a different isotopic composition is specified (i.e., unless the species is isotopically modified). The isotopic modification may arise because the ratio of isotopes present is significantly different than the natural ratio or because only a single isotope is present. Species of both types may be named by the system of nomenclature recommended by IUPAC for isotopically modified inorganic compounds (*26*). An isotopically modified organic group occurring in an inorganic compound, such as an organic ligand in a coordination entity, is named according to the procedure given in Section H of the 1979 IUPAC Organic Rules (*34p*).

Another general system for describing isotopically modified compounds is based on an extension of the principles proposed by Boughton for the nomenclature of compounds containing hydrogen isotopes (*76*). It is currently used mainly in the index nomenclature of CAS (*32d*), and it is also found in the literature and in many chemical catalogs. Because the system recommended by CNIC provides for recognition of various types of isotopic modifications, it will be used as the basis for the bulk of this chapter. The system based on the Boughton principles will be discussed at the end of this chapter (*see* Section 15.5).

Isotopically modified species may be classified into two broad categories, isotopically substituted species and isotopically labeled species. *Isotopically substituted species* are those in which essentially all individual molecules (or ions) contain the same isotope of an element at each designated atomic position (*see* Section 15.3). *Isotopically labeled species* are those in which the ratio of isotopes of an element at each designated atomic position is significantly different from the natural ratio. They may be considered formally to be a mixture of an isotopically unmodified species and one or more analogous isotopically substituted species (*see* Section 15.4). Isotopically labeled species may be further divided into specifically labeled, selectively labeled, nonselectively labeled, and isotopically deficient species. *Specifically labeled* means that both the atomic position(s) and the number of each labeling nuclide are defined. *Selectively labeled* means that the atomic position(s), but not necessarily the number, of each labeling nuclide is defined. *Nonselectively labeled* means that both the atomic position(s) and number of the labeling nuclide(s) are undefined. *Isotopically deficient* means that the natural isotopic content of one or more elements in the species has been depleted.

A formula written in the usual manner (*see* Section 2.4) implies that the compound is isotopically unmodified, as does a name based on the formula. Such a name requires alteration only if it is desired to contrast the "natural" nature of the compound to an isotopically modified compound or to emphasize the compound's "natural" or "normal" character. For example, $PH_3$ with natural isotopic compositions for phosphorus and hydrogen is called "unmodified phosphine".

## 15.2 Symbolism

The symbol used to denote a specific nuclide in a formula or name for an isotopically modified species consists of the atomic symbol for the element (*see* Table A.I) with an arabic number indicating the mass number of the nuclide as a left superscript (*see* Section 3.2.4) (example: $^{130}Xe$). A metastable nuclide is indicated by adding the letter "m" to the mass number of the nuclide (example: $^{133m}Xe$). The atomic symbol is printed in Roman type because italicized atomic symbols are used as locants in both coordination nomenclature (*see* Section 5.3.5.1) and organic chemical nomenclature (*34*). When it is necessary to cite different nuclides at the same place in the formula or name of an isotopically modified species, the nuclide symbols are written in alphabetical order according to their atomic symbols [example: ($^{2}H_2$,$^{35}S$)sulfuric acid]. When the atomic symbols are the same, the nuclide symbols are written in order of increasing mass number [example: ($^{78}Br$,$^{81}Br$)dibromine]. Nuclide symbols, with any locants, are separated from each other by commas. When both isotopically unmodified and modified forms of an element occur in equivalent atomic positions, the atomic symbol for the element with appropriate subscript precedes the nuclide symbol(s) for the element in the formula (example: $SiH_3{}^2H$).

For the hydrogen isotopes with mass numbers 1 (protium), 2 (deuterium), and 3 (tritium), the nuclide symbols $^1H$, $^2H$, and $^3H$, respectively, are preferred. The symbols D and T may be used for $^2H$ and $^3H$, respectively, but not when other labeling nuclides are present because then the use of D and T may lead to difficulties in alphabetical ordering of the nuclide symbols in the isotopic descriptor. The symbols *d* and *t* are used in place of $^2H$ and $^3H$ in names formed according to the Boughton system (*see* Section 15.5.1). In no other case are single lower-case letters used as atomic symbols, so the use of *d* and *t* outside of the Boughton system is not recommended.

## 15.3 Isotopically Substituted Species

### 15.3.1 Definitions and Formulas

An isotopically substituted species has a composition such that essentially all the molecules (or ions) of the species have only the indicated nuclide(s) at each designated atomic position. The absence of nuclide indication for any atomic position means that the isotopic composition is the natural one for that position. The formula for an isotopically substituted species is written as usual (*see* Section 2.4), except that nuclide symbols are used instead of atomic symbols where appropriate. *See* Section 15.3.2 for examples.

### 15.3.2 Names

The name of an isotopically substituted species is formed by enclosing the appropriate nuclide symbols with any necessary locants in parentheses and using the resulting descriptor as a prefix to the name of the species, or preferably to the name of the part of the species that is isotopically substituted. Immediately after the parentheses there is neither space nor hyphen, except that a hyphen is inserted when the name of the species or part begins with a locant. When isotopic polysubstitution is possible, the number of atoms that have been replaced is always specified as a right subscript to the atomic symbol in the nuclide symbol, even in the case of monosubstitution.

Examples:

| | |
|---|---|
| $H^{3}HO$ | ($^{3}H_{1}$)water |
| $^{78}Br^{81}Br$ | ($^{78}Br$,$^{81}Br$)bromine |
| $H_2{}^{35}SO_4$ | ($^{35}S$)sulfuric acid |
| $Na^{36}Cl$ | sodium ($^{36}Cl$)chloride |
| $^{235}UF_6$ | ($^{235}U$)uranium(VI) fluoride<br>($^{235}U$)uranium hexafluoride |
| $^{15}N_2O$ | di[($^{15}N$)nitrogen] oxide |
| $^{11}BH(OC^{2}H_3)_2$ | di[($^{2}H_3$)methoxy]($^{11}B$)borane |
| $^{113m}InCl_3$ | ($^{113m}In$)indium(3+) chloride<br>($^{113m}In$)indium(III) chloride<br>($^{113m}In$)indium trichloride |
| $H_2{}^{2}H\underset{1}{Si}\underset{2}{Si}H^{36}Cl\underset{3}{Si}H_3$ | 2-($^{36}Cl$)chloro(1-$^{2}H_1$)trisilane |
| $Ca_3({}^{32}PO_4)_2$ | tricalcium bis[($^{32}P$)phosphate] |
| $^{42}KNa^{14}CO_3$ | ($^{42}K$)potassium sodium ($^{14}C$)carbonate |
| $K[{}^{32}PF_6]$ | potassium hexafluoro($^{32}P$)phosphate |
| $[{}^{50}Cr({}^{2}H_2O)_6]Cl_3$ | hexa[($^{2}H_2$)aqua]($^{50}Cr$)chromium(3+) chloride<br>hexa[($^{2}H_2$)aqua]($^{50}Cr$)chromium(III) chloride |
| $[Mo({}^{15}N_2)_2(Ph_2PCH_2CH_2PPh_2)_2]$ | *trans*-{bis[($^{15}N_2$)dinitrogen]bis[ethylene-bis(diphenylphosphine)]molybdenum} |

When there is no ambiguity, nuclide symbols may be cited before a numerical prefix. This convention is followed mainly to retain the same name alphabetically for the isotopically substituted species as for the unmodified species insofar as possible. When there is ambiguity, such as when identical units in an unmodified structure are not identically modified, the individual units must be cited separately. Specific guidance on ordering of modified and unmodified identical groups in a name is lacking in current recommendations.

Examples:

| | |
|---|---|
| $B^{35}Cl^{37}Cl_2$ | boron ($^{35}Cl_1$,$^{37}Cl_2$)trichloride<br>($^{35}Cl_1$,$^{37}Cl_2$)trichloroborane |
| $K_3{}^{42}K[Fe(CN)_6]$ | ($^{42}K_1$)tetrapotassium hexacyanoferrate |
| $[(OC)_2Rh(\mu\text{-}{}^{35}Cl)(\mu\text{-}{}^{37}Cl)Rh(CO)_2]$ | tetracarbonyl[($^{35}Cl$,$^{37}Cl$)di-$\mu$-chloro]dirhodium |
| $(CH^{2}H_2)\underset{2}{Si}H_2\underset{1}{Si}(CH_3)HOH$ | 1,2-(2-$^{2}H_2$)dimethyl-1-disilanol |

| | |
|---|---|
| $OCH_3$<br>\|<br>$C_6H_5PO$<br>\|<br>$S^{14}CH_3$ | *O*-methyl *S*-($^{14}$C)methyl phenylphosphonothioate |

## 15.4 Isotopically Labeled Species

### 15.4.1 Specifically Labeled Species

*15.4.1.1 **Definitions and Formulas.*** A specifically labeled species (a species in which both the position(s) and the number of each labeling nuclide are defined) is formed conceptually by adding a unique isotopically substituted species to the corresponding isotopically unmodified species. It is singly labeled when the isotopically substituted species has only one isotopically modified atomic position, multiply labeled when the isotopically substituted species has isotopic modification of the same element at different positions or multiply at the same position, and mixed labeled when the isotopically substituted species has more than one kind of isotopically modified element.

The formula of a specifically labeled species is written in the usual way (*see* Section 2.4), except that the appropriate nuclide symbol(s) and multiplying subscript(s), if any, are enclosed in square brackets and substituted for the atomic symbols of the isotopically modified element(s). Although the formula for a specifically labeled species does not represent the isotopic composition of the bulk material, which is usually overwhelmingly that of the isotopically unmodified species, it does indicate the presence of the species of chief interest, the isotopically substituted species.

Examples:

| The addition of | to | yields |
|---|---|---|
| $H^{36}Cl$ | $HCl$ | $H[^{36}Cl]$ |
| $H^{99m}TcO_4$ | $HTcO_4$ | $H[^{99m}Tc]O_4$ |
| $^{32}PCl_3$ | $PCl_3$ | $[^{32}P]Cl_3$ |
| $Ge^2H_2F_2$ | $GeH_2F_2$ | $Ge[^2H_2]F_2$ |

*15.4.1.2 **Names.*** The name of a specifically labeled species may be formed by inserting the nuclide symbol(s), preceded by any necessary locant(s), in square brackets before the name of the corresponding isotopically unmodified species, or preferably before the name of the part that is isotopically modified. Immediately after the square brackets there is neither space nor hyphen, except that a hyphen is inserted when the name or part of a name requires a preceding locant. When it is possible to label more than one equivalent atomic position of the same element (e.g., $SiH_4$ has four equivalent H atomic positions), the number of atomic positions that have been labeled is always specified as a right subscript to the nuclide symbol, even when only one position is labeled. This convention is necessary in order to distinguish specific from selective or nonselective labeling (*see* Sections 15.4.2 and 15.4.3). The name of a specifically labeled species differs from that of the corresponding isotopically substituted species (*see* Section 15.3.2) only in the use of square brackets instead of parentheses around the nuclide descriptor.

Examples [(A) indicates singly labeled, (B) multiply labeled, (C) mixed labeled]:

| | |
|---|---|
| $H[^{36}Cl]$ | hydrogen [$^{36}$Cl]chloride (A) |
| $H[^{99m}Tc]O_4$ | [$^{99m}$Tc]pertechnetic acid (A)<br>hydrogen tetraoxo[$^{99m}$Tc]technetate(1 −) |

| | |
|---|---|
| $[^{32}P]Cl_3$ | $[^{32}P]$phosphorus trichloride (A) |
| $Ge[^{2}H_2]F_2$ | difluoro$[^{2}H_2]$germane (B) |
| $Na_2[^{35}S]O_3S$ | sodium thio$[^{35}S]$sulfate (A) |
| $Na_2SO_3[^{35}S]$ | sodium ($[^{35}S]$thio)sulfate (A) |
| $HO[^{18}O]H$ | hydrogen $[^{18}O_1]$peroxide (A) |
| $[^{15}N]H_2[^{2}H]$ | $[^{2}H_1,^{15}N]$ammonia (C) |
| $[^{13}C]O[^{17}O]$ | $[^{13}C]$carbon $[^{17}O_1]$dioxide (C) |
| $[^{32}P]O[^{18}F_3]$ | $[^{32}P]$phosphoryl $[^{18}F_3]$fluoride (C) |
| $H_3SiSiH[^{2}H_2]$<br>**2 1** | $[1,1\text{-}^{2}H_2]$disilane (B) |
| $[^{18}O]CH_3$<br>\|<br>$H[^{2}H]SiSiH_3$<br>**1 2** | 1-($[^{18}O]$methoxy)$[1\text{-}^{2}H_1]$disilane (C) |
| $H_2[^{10}B][^{2}H_2][^{10}B]H_2$ | $[^{10}B_2,\mu,\mu\text{-}^{2}H_2]$diborane(6) (C)<br>[The more complete ...,(1,2-$\mu$)-(1,2-$\mu$)-... is not needed in naming this simple borane.] |
| $[Fe(CO)_2([^{13}C]O)_2Br_2]$ | dibromo($[^{13}C_2]$tetracarbonyl)iron (B) |
| $[Al([^{13}C]H_3CO[^{13}C]HCO[^{13}C]H_3)_3]$ | tris($[1,3,5\text{-}^{13}C]$-2,4-pentanedionato)aluminum (B) |

### 15.4.2 Selectively Labeled Species

***15.4.2.1 Definitions and Formulas.*** A selectively labeled species (a species in which the atomic position(s), but not necessarily the number, of each labeling nuclide is determined) is formed conceptually by adding a mixture of isotopically substituted species to the corresponding isotopically unmodified species. A selectively labeled species may be considered to be a mixture of specifically labeled species. A selectively labeled species is termed *multiply labeled* when isotopic modification occurs at more than one atomic position, either in a set of equivalent atomic positions (example: H in $SiH_4$) or at different atomic positions in a molecule (example: B in $[B_8H_8]^{2-}$). It is termed *mixed labeled* when there is more than one labeling nuclide in the species (example: B and C in $B_8C_2H_{10}$). When there is only one atomic position for any element in a given species, only specific labeling (*see* Section 15.4.1) is possible.

A unique structural formula cannot be written for a selectively labeled species. Such a species may be represented by placing the nuclide symbol(s) preceded by the necessary locants, but without multiplying subscripts, enclosed in square brackets directly before the formula for the isotopically unmodified species (*see* Section 2.4). If necessary and possible, the insertion may be made before parts of the formula that have an independent numbering. Identical locants are not repeated. This method of designating selective labeling may also be used with molecular as well as structural formulas.

Examples:

| The addition of | to | yields |
|---|---|---|
| $SOCl^{36}Cl$ and $SO^{36}Cl_2$ | $SOCl_2$ | $[^{36}Cl]SOCl_2$ |
| $PH_2{}^{2}H$, $PH^{2}H_2$, and $P^{2}H_3$, or any two of these | $PH_3$ | $[^{2}H]PH_3$ |
| $H_2{}^{10}BH_2{}^{10}BHCl$, $H_2{}^{10}BH_2BHCl$, and $H_2BH_2{}^{10}BHCl$, or any two of these | $H_2BH_2BHCl$ | $[^{10}B]H_2BH_2BHCl$ |
| $H_3{}^{32}PO_4$, $H_3{}^{32}PO_3{}^{18}O$, $H_3{}^{32}PO_2{}^{18}O_2$, etc.; and $H_3PO_3{}^{18}O$, $H_3PO_2{}^{18}O_2$, etc., or any two or more of these with $^{32}P$ in at least one and P in a different one, and with $^{18}O$ in at least one and only O in a different one | $H_3PO_4$ | $[^{18}O,^{32}P]H_3PO_4$ |

*15.4.2.2 Names.* The name of a selectively labeled species may be formed by placing the nuclide symbol(s), preceded by any necessary locants, in square brackets before the name of the isotopically unmodified species, or preferably before the part of the name referring to the modified atomic position(s). Identical locants corresponding to atomic positions of the same element are not repeated. Immediately after the brackets there is neither space nor hyphen, except that a hyphen is inserted when a name or part of a name requires a preceding locant. The name of a selectively labeled species differs from the name of the corresponding isotopically substituted species in the use of square brackets rather than parentheses and in the omission of repeated identical locants and of multiplying subscripts, except as described in the paragraph that follows the examples.

Examples:

|  |  |
|---|---|
| $[^{36}Cl]SOCl_2$ | $[^{36}Cl]$sulfinyl chloride<br>(not $[^{36}Cl_2]$sulfinyl chloride) |
| $[^{18}O]H_3PO_4$ | $[^{18}O]$phosphoric acid<br>(not $[^{18}O_4]$phosphoric acid) |
| $[^{2}H]PH_3$ | $[^{2}H]$phosphine or $[^{2}H]$phosphane<br>(not $[^{2}H_3]$phosphine or $[^{2}H_3]$ phosphane) |
| $[^{10}B]H_2BH_2BHCl$ | chloro$[^{10}B]$diborane(6)<br>{not chloro$[^{10}B_2]$diborane(6)} |
| $[^{13}C]Fe(CO)_5$ | $[^{13}C]$pentacarbonyliron<br>(not $[^{13}C_5]$pentacarbonyliron) |
| $[^{15}N]K_3[Fe(CN)_6]$ | potassium $[^{15}N]$hexacyanoferrate(3 −)<br>{not potassium $[^{15}N_6]$hexacyanoferrate(3 −)} |

A selectively labeled species may be formed by conceptually mixing two or more, but not all possible, isotopically substituted species with the corresponding isotopically unmodified species. This situation may be designated by adding the number, or the possible number, of labeling nuclide(s) for each atomic position as subscripts to the nuclide symbol(s) in both the formula (*see* Section 15.4.2.1) and the name. Two or more subscripts referring to the same nuclide symbol are separated by a semicolon. The subscript zero is used to indicate that one of the isotopically substituted components is not modified at the indicated atomic position. For a multiply or mixed labeled species the subscripts are written successively in the same order as the various isotopically substituted species are considered.

Examples:

| The addition of | to | yields |
|---|---|---|
| $H_2{}^2HSiOSiH_2OSiH_3$ and<br>$H^2H_2SiOSiH_2OSiH_3$ | $H_3SiOSiH_2OSiH_3$ | $[1\text{-}^2H_{1;2}]H_2SiOSiH_2OSiH_3$<br>$[1\text{-}^2H_{1;2}]$trisiloxane |
| $H^2H_2SiOSiH_2OSiH_3$ and<br>$H^2H_2Si^{18}OSiH_2OSiH_3$ | $H_3SiOSiH_2OSiH_3$ | $[1,1\text{-}^2H_{2;2},2\text{-}^{18}O_{0;1}]H_3SiOSiH_2OSiH_3$<br>$[1,1\text{-}^2H_{2;2},2\text{-}^{18}O_{0;1}]$trisiloxane |
| $H_3Si^{18}OSiH_2OSiH_3$ and<br>$H^2H_2SiOSiH_2OSiH_3$ | $H_3SiOSiH_2OSiH_3$ | $[1\text{-}^2H_{0;2},2\text{-}^{18}O_{1;0}]H_3SiOSiH_2OSiH_3$<br>$[1\text{-}^2H_{0;2},2\text{-}^{18}O_{1;0}]$trisiloxane |

A selectively labeled species in which all atomic positions of a particular element are labeled in the same isotopic ratio may be indicated by adding the italicized descriptor "*unf*" (to indicate "uniform" labeling) immediately preceding, without hyphen, the nuclide symbol in the isotopic descriptor of the formula or name. The italicized descriptor *unf* may be followed by appropriate locants to indicate uniform labeling at specific positions.

Examples: [*unf*$^{13}$C]Fe(CO)$_5$ — [*unf*$^{13}$C]pentacarbonyliron

Comment: The excess $^{13}$C is equally distributed among the carbonyl ligands.

[*unf*$^{190}$Os]Os$_6$(CO)$_{18}$ — [*unf*$^{190}$Os]octadecacarbonylhexaosmium

Comment: The excess $^{190}$Os is equally distributed among all the osmium atomic positions.

[*unf*-1,5-$^{32}$Si]ClSiH$_2$OSiH$_2$OSiH$_3$ — 1-chloro[*unf*-1,5-$^{32}$Si]trisiloxane

Comment: The excess $^{32}$Si is equally distributed between the two terminal silicon atomic positions.

A selectively labeled species in which all the atomic positions for a particular element are modified isotopically, but not necessarily uniformly, may be indicated by adding the italicized descriptor "*gen*" (to indicate general labeling) immediately preceding, without hyphen, the nuclide symbol in the isotopic descriptor of the formula or name.

Examples: [*gen*$^{13}$C]Fe(CO)$_5$ — [*gen*$^{13}$C]pentacarbonyliron

Comment: The excess $^{13}$C is distributed among all the carbonyl ligands, but not necessarily uniformly.

[*gen*$^{190}$Os]Os$_6$(CO)$_{18}$ — [*gen*$^{190}$Os]octadecacarbonylhexaosmium

Comment: The excess $^{190}$Os is distributed among all the osmium atomic positions, but not necessarily uniformly.

### 15.4.3 Nonselectively Labeled Species

A nonselectively labeled species (a species in which neither the atomic position(s) nor the number of labeling nuclides is defined) is indicated in the formula and name by inserting the nuclide symbol(s), enclosed in square brackets, directly before the usual formula or name for the corresponding isotopically unmodified species. No preceding locants or subscripts are used.

Examples:

[$^{15}$N]HN$_3$ — hydrogen [$^{15}$N]azide

Comment: The excess $^{15}$N may occur at any of the three nitrogen atomic positions.

[$^{32}$P](HO)$_2$P(O)OP(O)(OH)OP(O)(OH)$_2$ — [$^{32}$P]triphosphoric acid

Comment: The excess $^{32}$P may occur at any of the phosphorus atomic positions.

[$^{18}$O]K$_4$H$_4$Si$_4$O$_{12}$ — tetrapotassium tetrahydrogen [$^{18}$O]*cyclo*-tetrasilicate

Comment: The excess $^{18}$O may occur at any of the oxygen atomic positions.

Isotopically labeled nonmolecular materials such as ionic solids and polymeric substances in which labeling nuclides may be randomly dispersed throughout a crystal lattice or polymeric molecule are considered to be nonselectively labeled and are given formulas and names with isotopic indicators appropriate for such labeling.

Examples: [$^{35}$Cl]NaCl — sodium [$^{35}$Cl]chloride

Comment: The excess $^{35}$Cl content is distributed randomly among the chloride ion sites in the crystalline solid.

[$^{235}$U]UO$_2$ — [$^{235}$U]uranium(IV) oxide

Comment: The excess $^{235}$U is distributed randomly among the uranium(IV) sites.

$[^{42}K]K_4H_4Si_4O_{12}$ [$^{42}$K]tetrapotassium tetrahydrogen *cyclo*-tetrasilicate
Comment: The excess $^{42}$K is distributed randomly among the potassium ion sites.

$[^{32}P](PNCl_2)_n$ *catena*-poly[(dichloro[$^{32}$P]phosphorus)-$\mu$-nitrido]
Comment: The excess $^{32}$P is distributed randomly among the phosphorus atomic positions.

## 15.4.4 Isotopically Deficient Species

Commercial products are available in which one or more isotopes of an element (particularly lithium, boron, carbon, nitrogen, uranium, and the Group 18 elements) have been partially or completely removed from a naturally occurring material. These products, therefore, do not have the "natural" isotopic composition for the element in question but are "isotopically deficient". In addition, certain other materials, such as meteorites, contain elements deficient in certain isotopes when compared to the so-called "natural" isotopic ratios. All of these isotopically deficient species may be designated by inserting the italicized morpheme "*def*", without a hyphen, immediately in front of the appropriate nuclide symbol and inserting the resultant descriptor enclosed in square brackets, without hyphens, in front of the formula or appropriate part of the name of the isotopically unmodified species.

Examples:

[*def*$^{10}$B]$H_3BO_3$ [*def*$^{10}$B]boric acid
Comment: The amount of $^{10}$B is less than the amount corresponding to the "natural" isotopic ratio for boron.

[*def*$^{235}$U]$UF_6$ [*def*$^{235}$U]uranium(VI) fluoride
Comment: The amount of $^{235}$U is less than the amount corresponding to the "natural" isotopic ratio for uranium.

## 15.4.5 Locants

*15.4.5.1* ***Basic Conventions.*** The name and numbering of an isotopically modified species should, whenever possible, be kept the same as that of the corresponding unmodified species by using locants for indicating positions of isotopic modification in an isotopically modified species that are part of the locant system normally used for the numbering of chains, rings, or clusters in the unmodified species. When there is a choice between equivalent chains or rings, the preferred chain or ring in the isotopically modified species is chosen so that it contains the maximum number of isotopically modified atoms or groups. If a choice still remains, precedence is given to the chain or ring that contains first a nuclide of higher atomic number and then a nuclide of higher mass number. However, extension of the Boughton principles (*see* Section 15.5) leads to assignment of lowest locants to isotopic positions in a "parent structure including unsaturation and principal groups, if any, before other considerations." Examples based on this procedure, which sometimes results in names and numbering different from those for the unmodified species, are given in Section 15.5.

Examples:

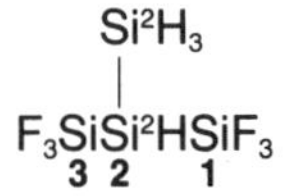

|  |  |
|---|---|
| $Si^{2}H_3$ attached at position 2 of $F_3SiSi^{2}HSiF_3$ (numbered 3 2 1) | 1,1,1,3,3,3-hexafluoro-2-[($^{2}H_3$)silyl](2-$^{2}H_1$)-trisilane |
| $[^{2}H_2]B_2$ (H bridges) $B_1$(Cl)(H) | 1-chloro[2,2-$^{2}H_2$]diborane(6) |
| ring: $O_1$, $B_2$(Cl), $O_3$, $[^{11}B]_4$(H), $O_5$, $HB_6$ | 2-chloro[4-$^{11}B_1$]boroxin |
| $SiHF_2$ attached at position 2 of $F_2[^{2}H_1]SiSiHSiF_3$ (numbered 3 2 1) | 2-(difluorosilyl)-1,1,1,3,3-pentafluoro[3-$^{2}H_1$]-trisilane {not 2-(difluoro[$^{2}H_1$]silyl-1,1,1,3,3-penta-fluorotrisilane; the trisilane chain containing $^{2}H$ is preferred} |
| $OSiH_2[^{2}H_1]$ attached at position 3 of $H_3Si[^{18}O]SiH[^{18}O]SiH_2Cl$ (numbered 5 4 3 2 1) | 1-chloro-3-([$^{2}H_1$]siloxy)[2,4-$^{18}O_2$]trisiloxane {not 1-chloro-3-([$^{18}O$]siloxy)[5-$^{2}H_1$,2-$^{18}O$]-trisiloxane; the trisiloxane chain containing both $^{18}O$ atoms is preferred} |

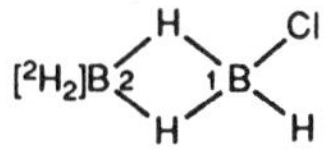

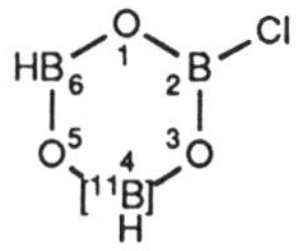

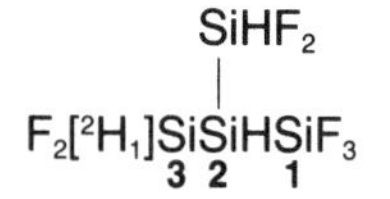

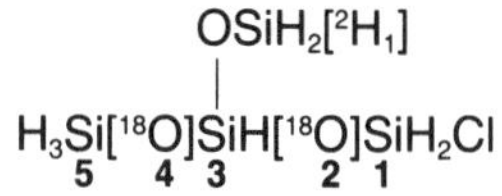

When there is a choice between equivalent numberings in an isotopically unmodified species, the starting point and direction of numbering for the corresponding isotopically modified species are chosen so as to give the lowest locants to the modified atoms or groups considered together as one series in ascending numerical order without regard to type of nuclide or mass number. If a choice still remains, preference for lowest locants is given first to a nuclide of higher atomic number and then to a nuclide of higher mass number.

Examples:

|  |  |
|---|---|
| $H_2[^{15}N]N[^{2}H_2]$ (numbered 2 1) | [1,1-$^{2}H_2$,2-$^{15}N$]hydrazine {not [2,2-$^{2}H_2$,1-$^{15}N$]hydrazine; the locant set 1,1,2 is preferred to 1,2,2} |
| $H_3SiSiH[^{2}H_1][^{29}Si]H_2SiH_3$ (numbered 4 3 2 1) | [3-$^{2}H_1$,2-$^{29}Si$]tetrasilane (not [2-$^{2}H_1$,3-$^{29}Si$]tetrasilane; Si is preferred to H) |
| ring: $O_1$, $O_2$, $[^{11}B]_3$(H), $O_4$, $H[^{10}B]_5$ | [5-$^{10}B$,3-$^{11}B$]-1,2,4,3,5-trioxadiborolane {not [3-$^{10}B$,5-$^{11}B$]-1,2,4,3,5-trioxadiborolane; $^{11}B$ is preferred to $^{10}B$} |
| $[^{29}Si]H_3$ attached at position 2 of $[^{2}H_3]SiN[^{30}Si]H_3$ (numbered 1 2 3) | 2-([$^{29}Si$]silyl)[1,1,1-$^{2}H_3$,3-$^{30}Si$]disilazane {not 2-([$^{29}Si$]silyl)[3,3,3-$^{2}H_3$,1-$^{30}Si$]disilazane; the locant set 1,1,1,3 is preferred to 1,3,3,3} {not 2-([$^{30}Si$]silyl)[1,1,1-$^{2}H_3$,3-$^{29}Si$]disilazane; $^{30}Si$ is preferred to $^{29}Si$ as a chain atom} |

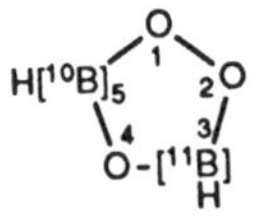

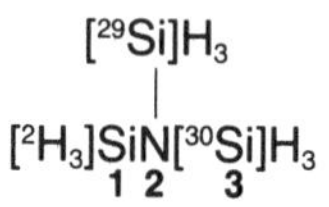

*15.4.5.2 **Additional Conventions.*** Isotopic modification at a position in a structure that is not normally assigned a locant may be indicated by a group symbol [i.e., the formula for the group with the nuclide symbol(s) for the isotopically modified atomic position(s)] or by attaching the italicized name for the group as a prefix to the nuclide symbol.

Examples: $HOSO_2[^{35}S]H$ — [$^{35}SH$]thiosulfuric acid; [*mercapto*-$^{35}S$]thiosulfuric acid

| | |
|---|---|
| $HO_3S[^{18}O][^{18}O]SO_3H$ | $[^{18}O^{18}O]$peroxodisulfuric acid<br>[*peroxo*-$^{18}O_2$]peroxodisulfuric acid |

Italicized nuclide symbols, capital italic letters, or both may be used as locants to distinguish between different nuclides of the same element.

Examples:

| | |
|---|---|
| $P(^{18}O)(OCH_3)_3$ | *O,O,O*-trimethyl ($^{18}O_1$)phosphate |
| $P(O)(OCH_3)_2[(^{18}O)CH_3]$ | *O,O,*$^{18}$*O*-trimethyl ($^{18}O_1$)phosphate |
| $(H_3C[^{18}O])SO_2(SCH_3)$ | $^{18}$*O,S*-dimethyl [$^{18}$O]thiosulfate |
| $[Ru(NH_3)_5(^{15}N^{14}N)]^{2+}$ | pentaammine[($^{14}$N,$^{15}$N)dinitrogen-$^{15}$*N*]ruthenium(2+) |

# 15.5 The CA System

## 15.5.1 The Boughton System

Organic compounds labeled with isotopes of hydrogen may be named according to the Boughton system (76) by placing the symbol for the isotope (with a subscript numeral to indicate the number of isotopic atoms) after the name or after the relevant portion of the name. The symbol for $^2$H (deuterium) is "*d*" and for $^3$H (tritium) is "*t*". No locant is needed if the isotopically modified atomic positions are equivalent or are completely substituted, but locants are cited if necessary. The locants (except Greek letters) and symbols are italicized, and hyphens are used to separate them from one another and from the remainder of the name.

Examples:

| | |
|---|---|
| $CH_3{}^2H$ | methane-*d* |
| $C^3H_4$ | methane-$t_4$ |
| $CH_3P^2H_2$ | methylphosphine-$d_2$ |
| $F_3\underset{2}{C}\underset{1}{C}H_2{}^2H$ | 2,2,2-trifluoroethane-*d*<br>(not 1,1,1-trifluoroethane-2-*d*; ethane-*d* is the parent) |
| $(^2H_2N)_2CO$ | urea-$d_4$ |

Fully deuterated or tritiated amines, imines, alcohols, phenols, thiols, selenols, and tellurols are named similarly, but partially labeled compounds of these classes and partially and fully deuterated organic acids have the "*d*" or "*t*" symbols placed immediately after the part(s) of the name to which they refer, with locants if necessary. This division is not performed with amides or with trivial names such as alanine.

Examples:

| | |
|---|---|
| $^2HON^2H_2$ | hydroxylamine-$d_3$ |
| $^2HONH^2H$ | hydroxyl-*d*-amine-*d* |
| $H_3SiN^2H_2$ | silanamine-$d_2$ |
| $^2H_3SiNH_2$ | silan-$d_3$-amine |
| $H_3CSO_3CH_2{}^3H$ | methyl-*t* methanesulfonate |

The "*d*" or "*t*" symbol is placed after the appropriate word in multiword names for classes not yet mentioned. In all other cases, including stereoparents, the isotopic symbol is cited after the complete name. Conventional locants or italicized words are often necessary to indicate the labeled atomic position. Labeled protonated species are indicated by terms such as "conjugate monoacid-*d*" and "monoprotonated-*d*".

Examples:

| | |
|---|---|
| $C^2H_3COO^2H$ | acetic-*d*$_3$ acid-*d* |
| $H_2{}^2HCCH_2CONH^2H$ (3 2 1) | propanamide-*N*,3-*d*$_2$ |
| $H^2H_2C$–(benzene ring)–$C^2HO$, $CH^2H_2$ at position 2 | 2,4-di(methyl-*d*$_2$)benzaldehyde-*formyl*-*d* |

Some other conventions used in CA index nomenclature for deuterated and tritiated organic compounds will now merely be mentioned. When locants are used in the name for a parent molecule or side group for unsaturation, hetero atoms, indicated hydrogen, spiro or ring-assembly junctions, bridges in fused ring systems, suffixes, or points of attachment (in side groups), locants are also cited for the labeled positions, whether or not their use would otherwise be necessary. Exceptions are "*d*" and "*t*" symbols placed after the word "acid" and after the suffixes of amines, imines, alcohols, etc., for which locants are seldom employed.

Enclosing marks are used with the name for a labeled side group if the name is preceded by a locant that designates its point of attachment to a parent molecule or to another side group. They are not used if the locant designates a position in the side group itself.

Formal addition of labeled hydrogen to a ring system is indicated by "hydro" substituents and the "*d*" or "*t*" symbol. When there is a choice, labeling is indicated in the name of the parent molecule rather than in the name of the side group.

For examples of the conventions outlined in the preceding three paragraphs, *see* the CA *Index Guide* (*32d*).

## 15.5.2 Organic Species with Non-Hydrogen Labeling

Isotopic modification of elements other than hydrogen in organic species is indicated by the appropriate symbols. Nomenclature is similar to that described in Section 15.5.1, except that the instances in which the isotopic symbol appears within the name are restricted to names consisting of more than one word; acids and acid derivatives with "carboxylic", "sulfonic", etc., names; and conjunctive names.

Examples:

| | |
|---|---|
| $CH_3C(^{17}O)(^{17}OH)$ | acetic-$^{17}$*O*$_2$ acid |
| $CH_3C(^{17}O)F$ | acetyl-$^{17}$*O* fluoride |
| $C(^{18}O)_2$ | carbon dioxide-$^{18}$*O*$_2$ |
| $H_3{}^{74}GeC^{15}N$ | germane-$^{74}$*Ge*-carbonitrile-$^{15}$*N* |
| $C_6H_5{}^{14}CH_2{}^{14}C(O)OH$ | benzeneacetic-*carboxy*,$\alpha$-$^{14}$*C*$_2$ acid |
| $H_3C$–(benzene ring, position 2 = $^{14}C$)–$^{35}SO_3H$ | benzene-2-$^{14}$*C*-sulfonic-$^{35}$*S* acid |

In general, for all other compounds isotopic labeling of a parent species is indicated after the name, although a combination of this convention with those for citing labeling with hydrogen isotopes is sometimes necessary. Symbols for isotopes of different elements that fall together are cited in alphabetical order and separated by hyphens. Labeling of side groups is indicated after the individual simple substituent group names.

Examples:

| | |
|---|---|
| $H_2{}^2HC^{14}CH_2OH$ (locants: 2 1) | ethan-*2-d*-ol-*1*-$^{14}$*C* |
| $CH_3{}^{13}CO^{15}NH_2$ (locants: 2 1) | acetamide-*1*-$^{13}$*C*-$^{15}$*N* |
| $CH_3CO^{15}N^2HCH_2CH^2HCH_3$ | N-(propyl-*2-d*)acetamide-*N-d*-$^{15}$*N* |
| $H_2{}^{14}C{=}CHCH_2O{-}C_6H_4{-}COOH$ (locants 3 2 1 on the propenyl chain; ring positions 1–6) | 4-(2-propenyl-*3*-$^{14}$*C*-oxy)benzoic acid |

Apart from isotopic modification of "hydro" ring species, multiple substituents that are identical except for labeling are named separately. Multiplicative nomenclature is not employed with unsymmetrically labeled parents. Instead, priority is based first on the maximum number of isotopically modified atomic positions and then on the alphabetically earliest isotope symbol.

Examples:

| | |
|---|---|
| $(CH_3)_2({}^{13}CH_3)_2Si$ | dimethyldi(methyl-$^{13}$*C*)silane |
| $^{35}Cl^{37}ClC$: | chloro-$^{35}$*Cl*-chloro-$^{37}$*Cl*-methylene |
| $(CH_3)_2{}^{14}CHNHC^2H(CH_3)_2$ | *N*-(1-methylethyl-*1-d*)-2-propanamine-*2*-$^{14}$*C* [not *N*-(1-methylethyl-*1*-$^{14}$*C*)-2-propan-*2-d*-amine; 2-propanamine-*2*-$^{14}$*C* is the parent] |

A name containing isotopic symbols is always used for a labeled species, regardless of how little of the labeled species is present, so long as its nature is known. When the number of labeled atomic positions is unknown, a phrase such as "labeled with deuterium" or "labeled with chlorine-37" is added to the name for the isotopically unmodified species. When two positions have equal degrees of labeling, the presence of isomers is indicated by a subscript, cited with the isotope symbol, that has a lower numerical value than the number of locants. For example, a 50–50 mixture of $C_6H_5{}^{14}CH_2COOH$ and $C_6H_5CH_2{}^{14}COOH$, in which the excess $^{14}C$ is present in two different atomic positions, would be named benzeneacetic-*carboxy*,α-$^{14}C_1$ acid.

### 15.5.3 Inorganic Species

Procedures similar to those in Sections 15.5.1 and 15.5.2 are used in naming isotopically modified inorganic species when the corresponding unmodified species has an unambiguous name. Labeled ligands in coordination compounds have the isotopic symbols appended to the ligand names, and the modified name is enclosed in parentheses if multiplicative prefixes are needed. Labeled hydro ligands are indicated by "*d*" or "*t*" symbols placed after citation of the hydro set.

Examples:

| | |
|---|---|
| $NH^2H^3H$ | ammonia-*d-t* |
| $(Si^2H_3)_3P$ | tri(silyl-$d_3$)phosphine |
| $^2HP(O)(OH)_2$ | phosphonic-*d* acid |
| $H_3{}^{32}PO_4$ | phosphoric-$^{32}$*P* acid |
| $^2H_2{}^{18}O$ | water-$d_2$-$^{18}$*O* |
| $^{22}Na_3PO_4$ | phosphoric acid tri(sodium-$^{22}$*Na*) salt<br>tri(sodium-$^{22}$*Na*) phosphate |
| $^3H^{35}Cl$ | hydrogen-*t* chloride-$^{35}$*Cl*<br>(aqueous solution: hydrochloric-$^{35}$*Cl* acid-*t*) |

| | |
|---|---|
| $[Al(^{2}H_2O)_6]^{3+}$ | hexa(aqua-$d_2$)aluminum(3+) |
| $[Co(NH_3)_5(N^{2}H_3)](ClO_4)_3$ | pentaammineammine-$d_3$-cobalt(3+) perchlorate |
| $H^{2}H_2AlN(CH_3)_2(CH_2{}^{2}H)$ | (*N,N*-dimethylmethan-*d*-amine)trihydro-$d_2$-aluminum |

When synonym line formulas are part of the designation for isotopically unmodified inorganic species, labeled analogs are identified by substituting isotope symbols for the appropriate atomic symbols in the synonym line formulas. Isotopic symbols are cited after unlabeled atomic symbols for the same element. Synonym line formulas containing isotopic symbols may also be cited for species that do not require synonym line formulas for the isotopically unmodified species.

Examples:

| | |
|---|---|
| $^{2}H_2S$ | hydrogen sulfide ($D_2S$) |
| $CoFe_2O_3{}^{18}O$ | cobalt iron oxide ($CoFe_2O_3{}^{18}O$)<br>cobalt(II) diiron(III) oxide ($CoFe_2O_3{}^{18}O$) |
| $Mo(CO)_4(^{13}CO)_2$ | molybdenum carbonyl ($Mo(CO)_4(^{13}CO)_2$)<br>molybdenum hexacarbonyl ($Mo(CO)_4(^{13}CO)_2$) |
| $Na_2{}^{35}S$ | sodium sulfide ($Na_2{}^{35}S$) |
| $^{231}UCl_3$ | uranium chloride ($^{231}UCl_3$)<br>uranium(III) chloride ($^{231}UCl_3$) |

Salts of inorganic oxo acids with isotopically modified cations may be identified by terms such as "strontium-$^{90}Sr$ salt" and "ammonium-$d_4$ salt" if no ratio is employed for the unlabeled salt. Otherwise the ratio is replaced by a labeled synonym line formula. Such formulas are also needed to indicate unsymmetrical labeling of anions.

Examples:

Phosphoric acid calcium salt ($Ca_2{}^{44}Ca(PO_4)_2$)
Phosphoric-$^{32}P$ acid calcium salt ($Ca_3(PO_4)(^{32}PO_4)$)

Species with indefinite labeled structures are indicated by modifying the names for the unlabeled species with "labeled with ... " phrases.

Isotopically modified alloys are indicated by including isotope symbols in the elemental composition included in each name (example: aluminum alloy, base, Al 95, $^{233}U$ 5). Synonym line formulas are cited instead of ratios for intermetallic compounds [example: compound of iron with uranium ($Fe_2{}^{235}U$)].

Labeled minerals are indicated by the addition of synonym line formulas that contain isotope symbols or phrases such as "labeled with boron-10" or "deuterated hydrate" to the name of the unlabeled mineral.

# Chapter 16
# Stereochemical Relationships

## 16.1 Introduction

Primarily because inorganic stereochemistry can be very complex, the development of methods to describe precisely the three-dimensional structures of inorganic species has been slow compared to the development of stereochemical nomenclature for organic compounds. The realization that atoms could form more than four bonds, leading to molecular arrangements other than tetrahedral, was its starting point, but the full scope of the subject was not apparent until complete structural determinations became possible. Most of the work on inorganic stereochemistry has been done with coordination compounds, so its discussion is usually intimately connected with the study of coordination chemistry. The general application of the conventions involved is, however, quite straightforward because stereochemistry is based on the number and spatial arrangement of bonds formed by a central atom, not the kinds of bonds involved. In order to generalize the discussion, it will be presented in terms of side groups, bonding atoms, and bonding number in place of the terms ligands, ligating atoms, and coordination number that have come to be associated with coordination compounds.

The relatively simple methods originally used to designate steric relationships in coordination compounds will be discussed first. The possibilities for isomerism as the bonding number of the central atom increases, however, quickly become so large that these methods are inadequate for distinguishing many individual isomers. For example, there are three possible *mer*-$a_3$ isomers for octahedral [$Ma_3bcd$] in which a, b, c, and d are monodentate side groups bonded to the central atom M. It thus becomes necessary to be able to designate the geometrical position of each side group attached to a central atom. Symbolism for this purpose is discussed in Section 16.3.

## 16.2 Simple Geometrical Designations

### 16.2.1 General Considerations

The discussion of spatial relationships in coordination compounds (*see* Section 5.3.6) lays the groundwork for the designation of the geometrical structure of all compounds with central atoms other than carbon. The use of simple prefixes to indicate the relationship of side groups to one another, as pointed out in that discussion, soon becomes inadequate for distinguishing among the various isomers possible in more complex situations. More comprehensive methods for locating side groups have been introduced from time to time, culminating in the procedure codified in the 1970 IUPAC Inorganic

Rules and outlined in Section 16.2.2. This procedure will probably be supplanted by the use of the stereochemical descriptors currently used by CAS (*32e*) and discussed in Section 16.3.

### 16.2.2 Locants for Side Groups

The use of simple prefixes as locants (*see* Section 5.3.6.3) is acceptable if there is no ambiguity. The more precise method of assigning a locant to each ligand was first accomplished by numbering each vertex of the polyhedron forming the framework for the structure. Later, italicized lower-case Roman letters were used for the locants pertaining to an atomic center to avoid confusion with the use of numbers to designate the location of specific atoms in a molecule or ion.

The procedure in the 1970 IUPAC Inorganic Rules (*22u*) is as follows:

> 7.514—The assignment of locant designators for other configurations around a coordination centre is based on locating planes of atoms perpendicular to a major axis in each configuration and assigning locants in a fixed manner in each successive plane. The actual procedure is: first, locate the highest (and longest in case of a choice) order axis of rotational symmetry; second, where the axis is not symmetrical, choose that end with a single atom (or smallest number of atoms) in the first plane to be numbered; third, locate the first plane of atoms (atom) to receive locants; fourth, orient the molecule so that the first position to receive a locant in the first plane with more than one atom is in the twelve o'clock position; fifth, assign locant designators to the axial position or to each coordinating position in the first plane, beginning at the 12 o'clock position and moving in a clockwise direction; sixth, from the first plane, move to the next position and continue assignments in the same manner, always returning to the 12 o'clock position or in the position nearest to it clockwise before assigning any locants in that plane; seventh, continue this operation until all positions are assigned.

Examples:

The names in the examples that follow illustrate the use of letter locants. Equally acceptable names may be generated by the procedures of Chapter 5 and of Section 16.3. In the older literature the locants may be 1, 2, 3, ..., instead of *a, b, c*, ....

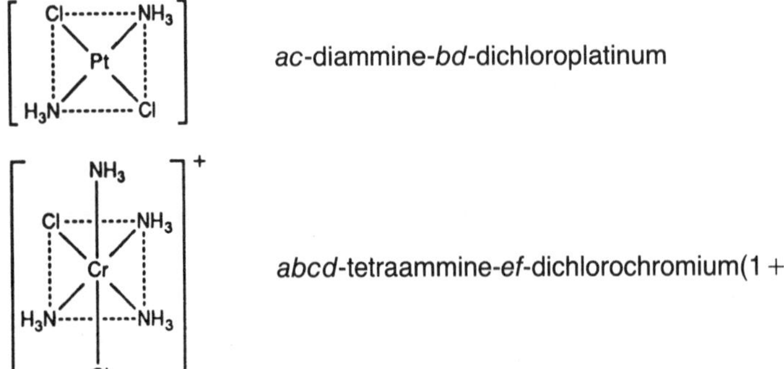

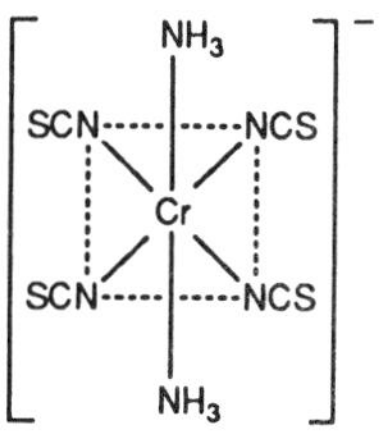

*af*-diammine-*bcde*-tetrakis(thiocyanato-k*N*)chromate(1 −)

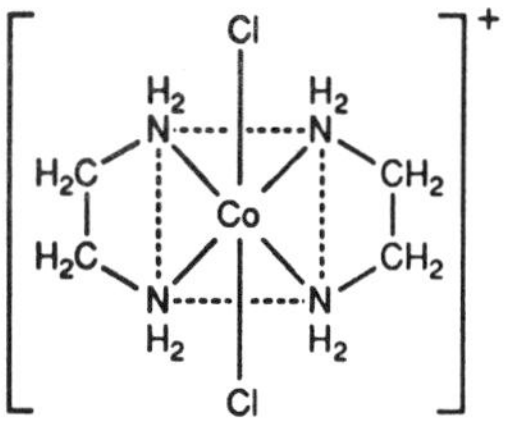

*af*-dichloro-*bc*,*de*-bis(1,2-ethanediamine-$k^2N,N'$)cobalt(1 +)

This system (*22u*) also included a method for indicating the geometrical configuration for each coordination number. For example, the two coordination number 5 configurations, the trigonal bipyramid and the square pyramid, were designated *5A* and *5B*, respectively, and the octahedron was designated *6A*. These arbitrary descriptors have not been generally adopted. Furthermore, the use of only the clockwise direction for locants, which does not allow the names for enantiomers to have the same locants but different descriptors, has also contributed to the failure of this system to be widely accepted.

## 16.3 Complete Stereochemical Descriptors

### 16.3.1 General Discussion

A stereonotation system developed by Brown, Cook, and Sloan (*77*) for use in the CAS chemical registry system and in CA index nomenclature may be used to designate the stereochemistry of the side groups around central atoms with bonding numbers from 2 to 9. It defines the geometrical arrangement of the side groups around the central atom, the relative geometric distribution of side groups within the structure, the chirality associated with the central atom, and the chirality associated with the side groups by means of a four-part descriptor. The four parts of the descriptor are given in the CA *Index Guide* (*32e*) as follows:

- a symbol indicating overall geometry called a "symmetry site term", which is called a "polyhedral symbol" in the 1990 IUPAC Inorganic Rules (*30*)
- a symbol called a "configuration number" indicating the arrangement of the side groups on each axis and plane of the polyhedron
- a symbol indicating the chirality associated with the central atom called a "chirality symbol"
- a symbol indicating the stereochemistry of each side group, when applicable, called the "ligand segment"

These four parts are enclosed in parentheses and separated from each other by hyphens to form the descriptor, which is affixed to the name of the species as a prefix separated by a hyphen. Examples are given in Section 16.4.

### 16.3.2 Symmetry Site Terms

Symmetry site terms consist of italicized one- to three-letter abbreviations describing the geometry, connected by a hyphen to an arabic numeral giving the bonding number of the central atom. The symmetry site terms for the common geometrical configurations based on bonding numbers from 2 to 9 are included as part of the various entries in Chart A.2.

### 16.3.3 Configuration Numbers with Monodentate Ligands

The configuration number, which is a shorthand notation for locating the side groups, is based on the order of seniority of the side groups as determined by using the CIP sequence rule (*35*). This ranking of the side groups depends first on the descending order of atomic number of the atoms directly attached to the central atom and then on the descending order of atomic mass number. Succeeding criteria involve stereochemical configurations in the side groups. The side groups are ranked by comparing them at each step in bond-by-bond exploration along the successive bonds of each side group by exhausting each criterion in turn until total ordering is achieved. When there is branching, atoms are compared in order of decreasing precedence of their paths.

The need for and meaning of a configuration number depend upon the site symmetry. Stereoisomerism is generally limited to geometrical differences for bonding numbers of 2 and 3. Thus the two side groups in a species with the bonding number of 2 can be on opposite sides of the central atom (linear) or they can form an angle of less than 180° with the central atom (angular). These two possibilities have been given the symmetry site terms *L*-2 and *A*-2, respectively. Configuration numbers are not needed for these species because neither of them can exist in isomeric forms dependent on the geometry around the central atom. The same situation prevails for the two geometrical configurations for species with a bonding number of 3, triangular (or planar) designated *TP*-3 and the so-called pyramidal designated *TPY*-3, and for one of the two configurations for species with a bonding number of 4, tetrahedral, designated *T*-4.

Geometrical isomerism is possible with the other configuration for a bonding number of 4, square (*see* Section 5.3.6.1 for a comment on "square planar"), designated *SP*-4, so its symmetry site term requires a configuration number to eliminate ambiguity. Up to three isomers are possible for a square configuration. They may be distinguished by citing the seniority ranking number of the bonding atom diagonally opposite the most senior bonding atom, which is considered to rank first, as the configuration number; for example,

has the configuration number 2 if the four side groups have the order of seniority shown. Because a planar structure cannot exhibit chirality, the complete stereochemical descriptor for such a configuration is (*SP*-4-2)-. The two isomers of this structure are designated (*SP*-4-3)- and (*SP*-4-4)-. These descriptors are to be used as prefixes in naming species with the respective patterns of bonding.

There are also two common configurations for species with a bonding number of 5. One is a trigonal bipyramidal arrangement of the side groups about the central atom, given the symbol *TB*-5. The configuration number is a two-digit number consisting of the seniority ranking numbers for the bonding atoms at the ends of the major axis cited in

descending order of priority. Thus,

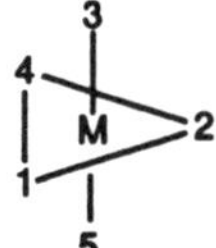


has the configuration number 35, if the order of seniority of the side groups is as shown. The nine isomers of this structure have the configuration numbers 12, 13, 14, 15, 23, 24, 25, 34, and 45. The stereochemical descriptors of the form (*TB*-5-35)- may be incomplete because the structure may be chiral. The designation of chirality is discussed in Section 16.3.6.

The second structure with a bonding number of 5 is square pyramidal and has the symmetry site term *SP*-5. The first digit of the two-digit configuration number for the distribution of side groups in the *SP*-5 symmetry is the seniority ranking number for the bonding atom on the principal axis. The second digit is the seniority ranking number for the bonding atom diagonally opposite the most senior bonding atom in the square base of the pyramid. For example, the configuration number for

is 34, if the seniority ranking for the side groups is as shown. There are 14 isomers of this structure with configuration numbers such as 13, 15, 23, 24, 25, 32, 43, and 51. Chirality is also possible for this structure, so that stereochemical descriptors of the form (*SP*-5-34)- may be incomplete if chirality needs to be specified (*see* Section 16.3.6).

Most species with a bonding number of 6 have an octahedral configuration about the central atom, although some have a trigonal prismatic configuration. The configuration number for an octahedral configuration consists of two digits. The first digit is the seniority ranking number of the bonding atom opposite the most senior bonding atom, and the second is the seniority ranking number of the bonding atom opposite the senior bonding atom in the plane perpendicular to the axis through M and the most senior bonding atom. These two digits are sufficient to specify the geometrical distribution of the side groups about the central atom, but chirality is again a possibility that may have to be designated if present. The octahedral structure

has the configuration number 64, if the order of seniority of its side groups is as shown. The partial stereochemical descriptor, including the symmetry site term and the configuration number, is then (*OC*-6-64)- for this distribution of side groups. There are 14 other possible distributions, among them those with configuration numbers of 63, 65, 53, 54, 56, 43, 45, and 46.

The less commonly encountered structure with a bonding number of 6, trigonal prismatic, has the symmetry site term *TP*-6. Configuration numbers for the distribution of bonding atoms in this symmetry are obtained by citing the seniority ranking numbers of the three bonding atoms in the triangular face opposite the preferred triangular face (i.e., the triangular face containing the maximum number of bonding atoms of highest seniority). These numbers are cited to correspond to the ascending numerical order of the seniority ranking numbers of the eclipsing bonding atoms in the preferred face. Thus, for the trigonal prismatic arrangement of the side groups

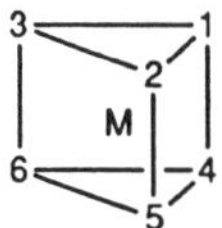


the configuration number would be 456, if the order of seniority is as shown. The resulting partial descriptor may still require a designation of chirality. Configuration numbers for some of the 59 possible isomers of this structure are 465, 654, 265, and 165.

The stereochemical descriptors for species with bonding numbers of 7, 8, and 9 are derived starting with the following symmetry site terms for the geometrical arrangements indicated:

| | |
|---|---|
| *PB*-7 | pentagonal bipyramidal |
| *OCF*-7 | mono(face-capped) octahedral |
| *TPS*-7 | mono(square-face-capped) trigonal prismatic |
| *CU*-8 | cubic |
| *SA*-8 | square antiprismatic |
| *DD*-8 | dodecahedral |
| *HB*-8 | hexagonal bipyramidal |
| *OCT*-8 | *trans*-bicapped octahedral |
| *TPT*-8 | bis(triangular-face-capped) trigonal prismatic |
| *TPS*-8 | bis(square-face-capped) trigonal prismatic |
| *TPS*-9 | tris(square-face-capped) trigonal prismatic |
| *HB*-9 | heptagonal bipyramidal |

The configuration number is then assigned by orienting the model structure with the highest order axis in the vertical plane and viewing it from the more senior bonding atom on the highest order axis or from a point on the highest order axis above the more preferred terminal (or end) plane perpendicular to this axis. The more preferred terminal (or end) plane is the end plane that, first, contains the greatest number of bonding atoms; second, contains the greatest number of highest seniority bonding atoms; or, third, is adjacent to a plane containing the greatest number of highest seniority bonding atoms.

The configuration numbers for those model structures with bonding atoms on the highest order axis (i.e., *PB*-7, *OCF*-7, *TPS*-7, *HB*-8, *OCT*-8, *TPT*-8, and *HB*-9) begin with the seniority ranking number(s) for the two bonding atoms on the highest order axis in lowest-numerical-order sequence. These digits are separated from the remainder of the configuration number by a hyphen. The remaining portion of the configuration number is derived by

- viewing the structure from the more senior bonding atom on the highest order axis or from the axial bonding atom located above the preferred plane,
- citing the seniority ranking number of the most senior bonding atom in the plane adjacent to that atom, and
- continuing to cite the seniority ranking numbers of the bonding atoms in sequence as they are encountered in the projection of the model structure, alternating between planes as necessary, in the direction that gives the lower numerical sequence.

When a bonding atom eclipses one or more bonding atoms in the projection of the model structure, all are cited, with the ranking number for the atom in the first plane preceding the ranking number(s) of the eclipsed atom(s), before continuing on to citation of the next bonding atom in the projection. For species in which clockwise and counterclockwise citations give two different configuration numbers (generally chiral species), the correct number is the one with the lower digit at the first point of difference. Thus the hexagonal bipyramidal structure

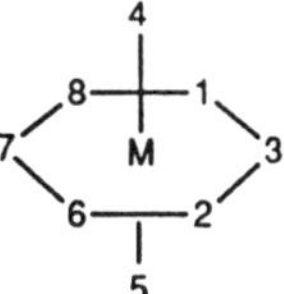


with the order of seniority of side groups as shown has the configuration number 45-132678 and the partial stereochemical descriptor *HB*-8-45-132678.

The remaining five configurations (i.e., *CU*-8, *SA*-8, *DD*-8, *TPS*-8, and *TPS*-9), which do not have a bonding atom on the highest order axis, do not have a distinct portion of the configuration number set off by a hyphen. These model structures are viewed from a point on the highest order symmetry axis above the preferred terminal plane. The configuration number is derived by citing first the seniority ranking number of the most senior bonding atom in the preferred terminal plane and then citing the seniority ranking number of the atom(s), if any, that it eclipses. The seniority ranking numbers of the remaining bonding atoms are then cited in order as they are encountered by proceeding either clockwise or counterclockwise around the projection of the model structure, alternating between planes as necessary and citing eclipsed bonding atoms immediately after the bonding atoms eclipsing them. Again, the correct configuration number is the one with the lower digit at the first point of difference when citation in clockwise and counterclockwise directions gives different results. For example, the cubic structure

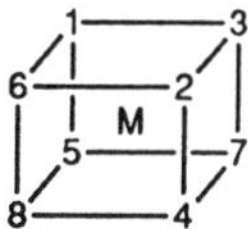


with the order of seniority of the side groups shown has the configuration number 15372468 and the partial stereochemical descriptor *CU*-8-15372468.

Chart A.2 gives typical stereochemical descriptors, which perforce contain examples of configuration numbers associated with each symmetry site term.

### 16.3.4 Equivalent Monodentate Side Groups

In many species several side groups are identical and will consequently be assigned the same (or degenerate) seniority ranking numbers. In order to use such ranking numbers and still have an unambiguous stereochemical descriptor, it is necessary to introduce an additional convention called "the trans maximum difference rule", which is applicable to bonding numbers 4 through 6. This rule may be stated in the following way. Whenever a choice exists in distinguishing between constitutionally equivalent bonding atoms, preference is given to the bonding atom trans to the least senior bonding atom (the bonding atom with the highest seniority ranking number). For example, in octahedral $Ma_2b_2c_2$, if the seniority ranking of the side groups is a > b > c, the seniority ranking numbers are 1, 1, 2, 2, 3, and 3. The most senior side group chosen to be on the fourfold axis would be the "a" opposite "c", not the "a" opposite "b", in the isomers offering such a choice, and the partial descriptor would be (*OC*-6-32)-, not (*OC*-6-23)-. Unique configuration numbers of no more than three digits result from the use of this rule.

### 16.3.5 Polydentate Ligands

Seniority ranking numbers are assigned to the bonding atoms of polydentate side groups in the same way as for monodentate side groups. When there are two or more equivalent bidentate or tridentate side groups, their seniority ranking numbers are distinguished by primes, double primes, etc. In the cases of symmetrical quadridentate, quinquedentate, and sexidentate side groups, equivalent bonding atoms are distinguished by priming bonding atom priority numbers in half of the ligand, thereby reducing the polydentate side group to groupings of equivalent bidentate or tridentate side groups. An unprimed bonding atom is senior to an equivalent primed atom, which in turn is senior to an equivalent double-primed bonding atom, etc. (i.e., the seniority ranking number $n$ is preferred to $n'$). Primes are ignored, however, when the bonding atoms are not equivalent (e.g., the seniority ranking number $2'$ is preferred to 3).

### 16.3.6 Chirality Descriptors

Three different sets of symbols are employed to designate the chirality of a structure. Although these symbols all convey essentially the same information (*R*, *C*, and $\Delta$ indicate right-handed or clockwise; *S*, *A*, and $\Lambda$ indicate left-handed or counterclockwise), the principles defining right- and left-handed are different for each set of symbols. There is consequently no exact translation of *R* to *C* or *C* to $\Delta$, and the three different sets of symbols have been retained.

The chirality symbols *R* for a right-handed path and *S* for a left-handed path are used for tetrahedral bonding patterns for other elements in the same way that they are used in organic chemistry for tetrahedral carbon centers (*35*). The model structure is oriented so that the face toward the viewer is opposite the bonding atom with the lowest seniority. When the seniority ranking sequence of the three facial bonding atoms is clockwise, the chirality symbol is *R*; when counterclockwise, *S*. If the seniority ranking of the side groups for the tetrahedral species Mabcd is a > b > c > d, the designations for the faces opposite d, that is,

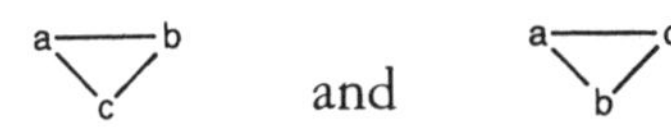

become *R* and *S*, respectively. The same descriptors are used for chiral centers with a bonding number of 3 with the so-called pyramidal structure. The pyramid is oriented so that the basal plane of the pyramid faces toward the viewer, with the central atom behind (or below) it.

The chirality of any octahedral species with two bidentate side groups in a "cis" relationship may be indicated by Δ or Λ, descriptors that can be related to right- and left-handed helixes, respectively (*22v*). To assign these descriptors it is necessary to relate two skew lines that are not orthogonal to each other to the edges of an octahedron. Two skew lines have one and only one normal in common. The projection of one line onto the other viewed along this common normal is either

B A-----A B or B A-----A B

where AA lies below BB. The two edges of an octahedron occupied by bidentate side groups in a cis relationship can be viewed as having the same relationship, that is,

B A–|–A M c–B d and B A–|–A M B–c d

in which AA and BB represent bidentate side groups that correspond, respectively, to the two skew-line projections just shown. The relationship illustrated by the sketches on the left, in which the direction from the lower line through the acute angle to the upper line in the projection is clockwise, is designated Δ. The relationship illustrated by the sketches on the right, in which the direction from the lower line through the acute angle to the upper line is counterclockwise, is then Λ. The descriptor derived for a given configuration is the same regardless of which bidentate ligand is chosen to define the bottom line, so seniority does not have to be established.

The chirality symbols used to denote absolute stereochemistry in all other inorganic structures are *C* for clockwise and *A* for counterclockwise. (*A* is derived from the less common term anticlockwise, which is used sometimes instead of the more familiar counterclockwise.) The structure is viewed either from the bonding atom of highest seniority on a major symmetry axis or from a point on a major axis above the preferred facial plane of the structure (*see* Section 16.3.3). As the projection is viewed from this point, the symbol *C* is used to signify that the direction from the most senior bonding atom to the more senior of the two bonding atoms adjacent to it in the projection is clockwise, ignoring any eclipsed atoms. Should the two neighboring atoms be equivalent, the next nearer bonding atoms are compared, etc., until a decision can be made. The symbol *A* signifies that the direction as described is counterclockwise.

Although the chirality associated with a side-group atom may be indicated in the side-group portion of the name, its indication in a fourth part of the stereochemical descriptor for the structure is preferred (*34q*). When more than one side group exhibits chirality, the chirality symbol(s) and/or term(s) for each are enclosed in separate sets of parentheses and cited in the order in which the side groups are cited in the name.

## 16.4 Examples

The development of names giving the complete stereochemical structures for a number of different species from their structural formulas (with the aid of planar projections where appropriate) follow. Unenclosed atomic symbols in the projections correspond to atoms in the plane of the paper. Symbols enclosed in one set of parentheses correspond to atoms lying beneath the plane of the paper, those enclosed in double sets of parentheses to atoms lying lower than those enclosed in one set, etc. Symbols enclosed in square brackets correspond to atoms lying above the plane of the paper. An enclosed atomic symbol touching another atomic symbol indicates that the atom in question lies directly below or above the other atom. Basic stereochemical descriptors for known structures with bonding numbers from 2 to 9 are listed in Chart A.2.

| Species | Name |
|---|---|
| [(pyridine)N—Au—Cl] | (*L*-2)-chloro(pyridine)gold(I) |

Comment: Only the symmetry site term is required for a complete description of the structure.

| Species | Name |
|---|---|
| Cl—S—Cl | (*A*-2)-sulfur dichloride |

Comment: Only the symmetry site term is required for a complete description of the structure.

| Species | Projection | Name |
|---|---|---|
| H, Cl—P, Br | H, (P)—Cl, Br | (*TPY*-3-S)-bromochloro-phosphine |

Comments: The trigonal pyramid is oriented so that the side groups are in the plane of the paper and the central atom is beneath it. The direction in the projection is counterclockwise from Br, the most senior bonding atom, to Cl, the next most senior bonding atom.

| Species | Projection | Name |
|---|---|---|
| F, H—Ge—Cl, Br | F, Cl—(Ge)((H)), Br | (*T*-4-*R*)-bromochloro-fluorogermane |

Comments: The order of seniority is Br > Cl > F > H. The tetrahedron is oriented so that H, the least senior bonding atom, lies below the plane of the paper, where it is eclipsed by the Ge in the projection. The direction in the projection is clockwise from Br, the most senior bonding atom, to Cl, the next most senior bonding atom.

| Species | Projection[a] | Name |
|---|---|---|
| [$(C_6H_5)_3P$, CO, OC—Fe, CO, $(C_6H_5)_3P$] | C, C—[P]Fe(P), C | (*TB*-5-11)-tricarbonylbis(tri-phenylphosphine)iron |

Comments: The order of seniority is P > C. The trigonal bipyramid is viewed from either axial bonding atom (P). The top P eclipses the Fe, which in turn eclipses the bottom P, in the projection. There is no chirality because the side groups in the equatorial plane are equivalent.

[a]Projection shows central and bonding atoms only.

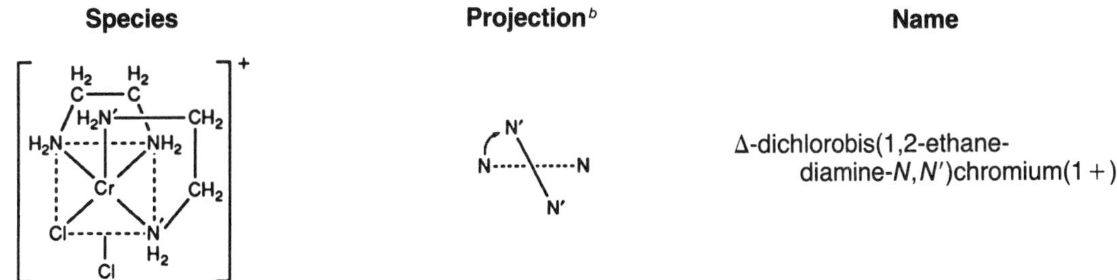


Comment: The direction from the lower edge through the acute angle to the upper edge is clockwise.

[b]Only the bidentate ligands (represented by N-----N and N′——N′) are shown.

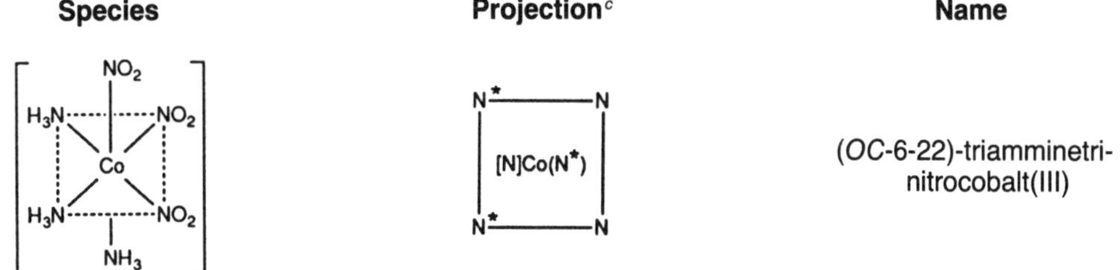


Comments: The order of seniority is N > N*. The octahedron is oriented so that it is viewed from $NO_2$. The N of this $NO_2$ eclipses the Co, which in turn eclipses the N of an $NH_3$. The structure does not exhibit chirality. This isomer is also known as the fac isomer. The mer isomer would have the descriptor (*OC*-6-12).

[c]N represents $NO_2$ and N* represents $NH_3$.

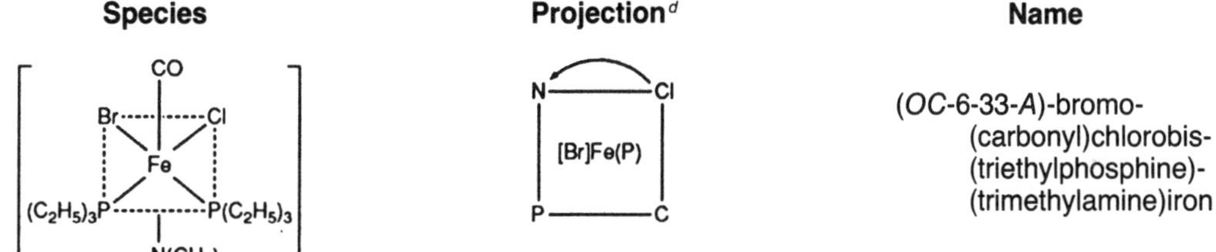


Comments: The order of seniority is Br > Cl > P > N > C. The octahedron is viewed from the Br, the most senior bonding atom. The Br eclipses the Fe, which in turn eclipses a P. The direction in the projection is counterclockwise from Cl, the most senior bonding atom in the plane, to N, its next most senior neighbor.

[d]Only the central and bonding atoms are shown.

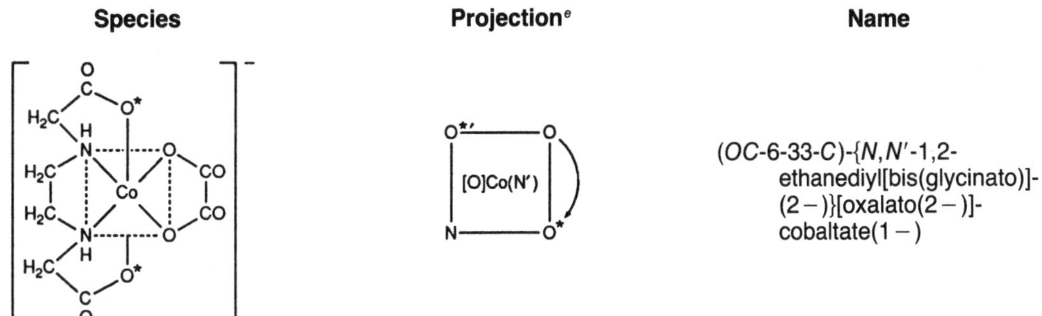


Comments: The order of seniority is O > O* > N. The octahedron is oriented so that it is viewed from an O. That O eclipses the Co, which in turn eclipses an N. Half the quadridentate side group is primed for ordering purposes. The direction in the projection is then clockwise from the O, the most senior bonding atom in the plane, to the unprimed O*, the next most senior.

[e]Only the central and bonding atoms are shown. O is from oxalate; O* is from glycinate.

| Species | Projection[f] | Name |
|---|---|---|
| | | (*TP*-6-11′1″)-tris[(2,2′-oxydiacetato)(2−)-$O^1$,$O^{1'}$]holmate(3−) |

Comments: The bonding O atoms of the bidentate side groups are primed for ordering purposes. The trigonal prism is viewed from a point on its threefold axis looking from above toward either one of the identical triangular faces. One bonding O atom from each side group eclipses the other bonding O atom from the same side group. The direction in the projection is from O, the most senior bonding atom, to O′, which is more senior than O″. The eclipsed atom has no effect on the choice of direction. There is no chirality.

[f]Only the central and bonding atoms are shown. The bonding atoms of the three bidentate ligands are differentiated by priming.

| Species | Projection | Name |
|---|---|---|

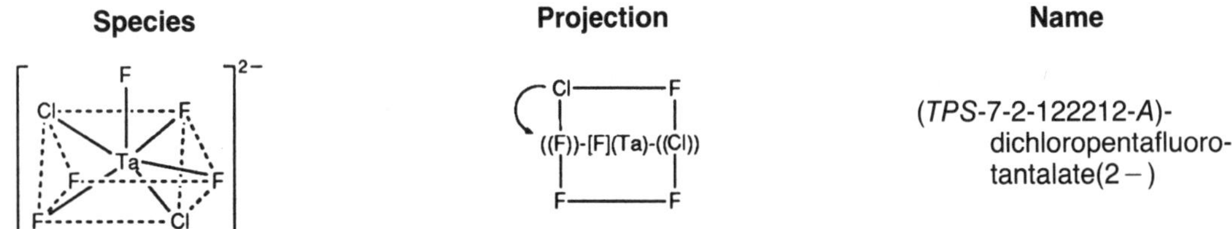


(*TPS*-7-2-122212-*A*)-dichloropentafluorotantalate(2−)

Comments: The order of seniority is Cl > F. The mono-square-face-capped trigonal prism is viewed from the F that is on the twofold axis and eclipses the Ta. The direction is counterclockwise from the Cl, the more senior atom in the top square face, toward the F in the lower plane of side groups in the projection. The alternate stereochemical descriptor (TPS-*7-2-121222-C*)-, which has a lower configuration number, is not preferred because the stipulation "alternating between planes where necessary" (*77b*) is interpreted to mean that moving in the direction that leads to the alternation of planes takes precedence over moving in the direction that yields the lower configuration number.

| Species | Projection[g] | Name |
|---|---|---|

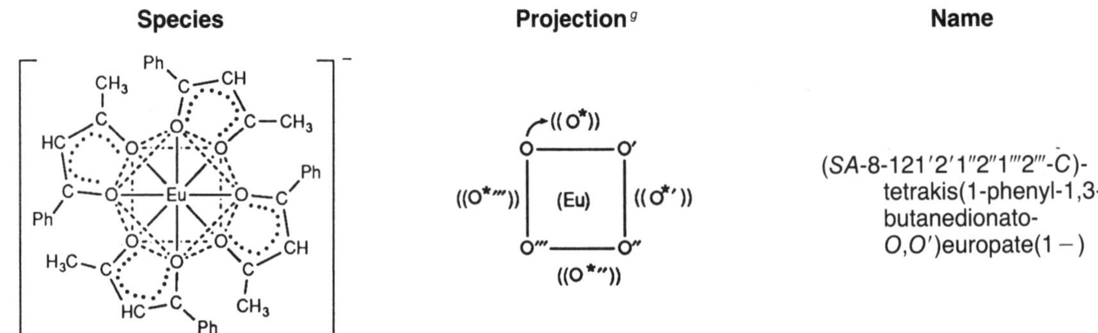


(*SA*-8-121′2′1″2″1‴2‴-*C*)-tetrakis(1-phenyl-1,3-butanedionato-*O*,*O*′)europate(1−)

Comments: The order of seniority is O > O*. The square antiprism is viewed from a point on its fourfold axis looking from above toward the square face formed by the four O atoms, which with their associated O* atoms are distinguished for ordering purposes by priming, etc. The direction in the projection is clockwise from O, the most senior atom in the top plane, to ((O*)), which is more senior than ((O*‴)).

[g]Only the central and bonding atoms are shown. O represents the O on the C next to the $C_6H_5$; O* represents the O on the C next to the $CH_3$.

| Species | Projection[h] | Name |
|---|---|---|
| | | (*OCT*-8-55-112346-*A*)-ammineaquadibromo-chlorofluorodioxouranate(1 −) |

Comments: The order of seniority is Br > Cl > F > O > O* > N. The *trans*-bicapped octahedron is viewed from the O* above the plane containing the more senior atoms Br, Cl, and O (in preference to Br, F, and N). The direction in the projection is counterclockwise from Br, the most senior atom in the top plane, to ((Br)), which is more senior than ((N)).

[h]O represents an O of $H_2O$, M* represents [O*]U(O*) if O* is the O bonded to U, and N represents $NH_3$.

| Species | Projection[i] | Name |
|---|---|---|

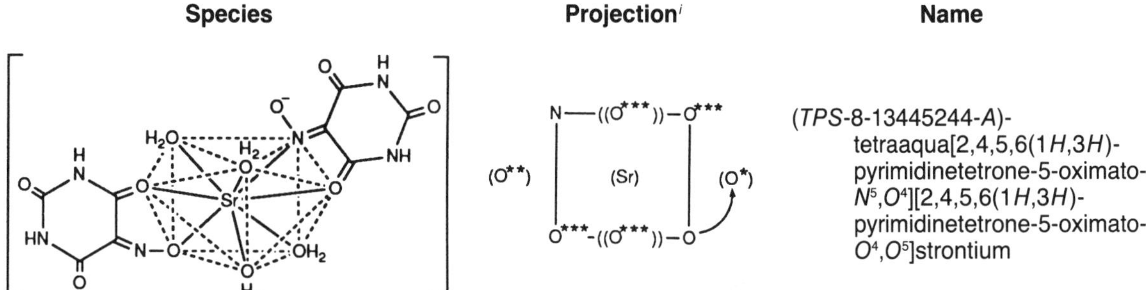


(*TPS*-8-13445244-*A*)-tetraaqua[2,4,5,6(1*H*,3*H*)-pyrimidinetetrone-5-oximato-$N^5,O^4$][2,4,5,6(1*H*,3*H*)-pyrimidinetetrone-5-oximato-$O^4,O^5$]strontium

Comments: The order of seniority is O > O** > O* > O*** > N. The bis(square-face-capped) trigonal prism is viewed from the twofold axis toward the uncapped square face. The direction in the projection is counterclockwise from O, the most senior bonding atom in the top plane to (O*), which is more senior than ((O***)).

[i]Only the bonding atoms are shown. O represents the O of the =NO and O* the bonding =O in the O,O-bonded $C_4H_2N_3O_4$; O** represents the bonding =O in the N,O-bonded $C_4H_2N_3O_4$; and O*** represents O in $H_2O$.

| Species | Projection[j] | Name |
|---|---|---|

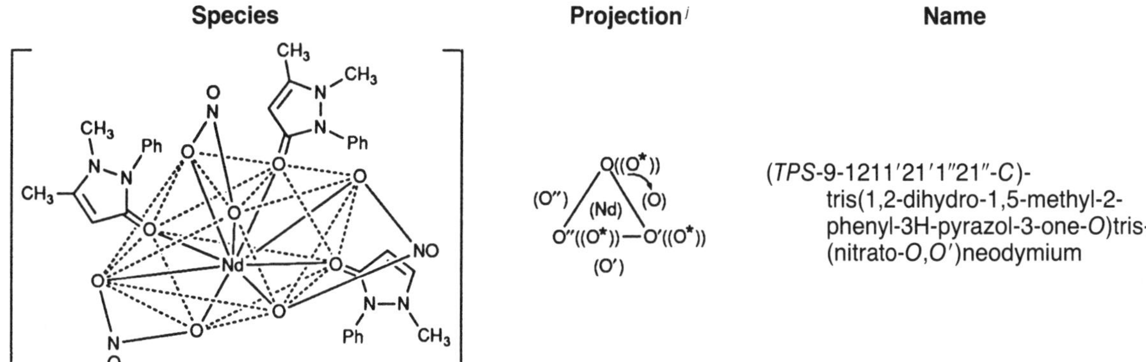


(*TPS*-9-1211′21′1″21″-*C*)-tris(1,2-dihydro-1,5-methyl-2-phenyl-3H-pyrazol-3-one-*O*)tris-(nitrato-*O*,*O*′)neodymium

Comments: The order of seniority is O > O*. Bidentate $NO_3$ side groups are primed, etc., for ordering purposes. The tris(square-face-capped) triangular prism is viewed from the threefold axis toward the triangular face formed by one O from each $NO_3$. Each facial O ($NO_3$) eclipses an O* ($C_{11}H_{12}N_2O$). The direction in the projection is clockwise from O, the most senior bonding atom in the top plane to (O), which is more senior than (O″). The eclipsed atom does not affect the choice of direction.

[j]Only the central and bonding atoms are shown. O represents the O in $NO_3$, and O* represents an O in $C_{11}H_{12}N_2O$.

# APPENDIX

**Table A.I. Names, Stems, Symbols, and Atomic Numbers for the Chemical Elements**

| *Name* | *Symbol* | *Atomic Number* |
|---|---|---|
| Actin-ium | Ac | 89 |
| Al/umin-um | Al | 13 |
| Americ/i-um | Am | 95 |
| Antimon-y (Stib-ium) | Sb | 51 |
| Argon- | Ar | 18 |
| Ars/en/i-c | As | 33 |
| Astat-ine | At | 85 |
| | | |
| Bar-ium | Ba | 56 |
| Berkel-ium | Bk | 97 |
| Beryll-ium | Be | 4 |
| Bism/uth- | Bi | 83 |
| Bor-on | B | 5 |
| Brom-ine | Br | 35 |
| | | |
| Cadm-ium | Cd | 48 |
| Calc/i-um | Ca | 20 |
| Californ-ium | Cf | 98 |
| Carb/on- | C | 6 |
| Cer-ium | Ce | 58 |
| Ces-ium | Cs | 55 |
| Chlor-ine | Cl | 17 |
| Chrom-ium | Cr | 24 |
| Cobalt- | Co | 27 |
| Copper (Cupr-um) | Cu | 29 |
| Cur-ium | Cm | 96 |
| | | |
| Dyspros-ium | Dy | 66 |
| | | |
| Einstein-ium | Es | 99 |
| Erb-ium | Er | 68 |
| Europ-ium | Eu | 63 |
| | | |
| Ferm-ium | Fm | 100 |
| Fluor-ine | F | 9 |
| Franc/i-um | Fr | 87 |
| | | |
| Gadolin-ium | Gd | 64 |
| Gall-ium | Ga | 31 |
| Germ/an-ium | Ge | 32 |
| Gold (Aur-um) | Au | 79 |
| | | |
| Hafn-ium | Hf | 72 |
| Hel-ium | He | 2 |
| Holm-ium | Ho | 67 |
| Hydr-ogen | H | 1 |
| Deuter-ium | D | |
| Trit-ium | T | |
| | | |
| Ind-ium | In | 49 |
| Iod-ine | I | 53 |
| Irid-ium | Ir | 77 |
| Iron (Ferr-um) | Fe | 26 |
| | | |
| Krypt-on | Kr | 36 |
| | | |
| Lanthan-um | La | 57 |
| Lawrenc/i-um | Lr | 103 |
| Lead (Plumb-um) | Pb | 82 |

Continued on next page.

**Table A.I.—Continued**

| *Name* | *Symbol* | *Atomic Number* |
|---|---|---|
| Lith-ium | Li | 3 |
| Lutet-ium | Lu | 71 |
| | | |
| Magnes-ium | Mg | 12 |
| Mangan-ese | Mn | 25 |
| Mendelev-ium | Md | 101 |
| Mercur-y (Hydrargyrum) | Hg | 80 |
| Molybd/en-um | Mo | 42 |
| | | |
| Neodym-ium | Nd | 60 |
| Neon- | Ne | 10 |
| Neptun-ium | Np | 93 |
| Nickel- | Ni | 28 |
| Niob-ium | Nb | 41 |
| Nitr-ogen | N | 7 |
| Nobel-ium | No | 102 |
| | | |
| Osm-ium | Os | 76 |
| Ox-ygen | O | 8 |
| | | |
| Pallad-ium | Pd | 46 |
| Phosph/or-us | P | 15 |
| Platin-um | Pt | 78 |
| Pluton-ium | Pu | 94 |
| Pol/on-ium | Po | 84 |
| Potass-ium (Kal-ium) | K | 19 |
| Praseodym-ium | Pr | 59 |
| Prometh-ium | Pm | 61 |
| Protactin-ium | Pa | 91 |
| | | |
| Rad-ium | Ra | 88 |
| Radon- | Rn | 86 |
| Rhen-ium | Re | 75 |
| Rhod-ium | Rh | 45 |
| Rubid-ium | Rb | 37 |
| Ruthen-ium | Ru | 44 |
| | | |
| Samar-ium | Sm | 62 |
| Scand-ium | Sc | 21 |
| Sel/en/i-um | Se | 34 |
| Sil/ic-on | Si | 14 |
| Silver (Argent-um) | Ag | 47 |
| Sod-ium (Natr-ium) | Na | 11 |
| Stront-ium | Sr | 38 |
| Sulf/ur- (Thi-on) | S | 16 |
| | | |
| Tantal-um | Ta | 73 |
| Technet-ium | Tc | 43 |
| Tell/ur-ium | Te | 52 |
| Terb-ium | Tb | 65 |
| Thall-ium | Tl | 81 |
| Thor-ium | Th | 90 |
| Thul-ium | Tm | 69 |
| Tin (Stann-um) | Sn | 50 |
| Titan-ium | Ti | 22 |
| Tungst/en- (Wolfram-) | W | 74 |
| | | |
| Unnilenn-ium | Une | 109 |
| Unnilhex-ium | Unh | 106 |
| Unniloct-ium | Uno | 108 |

**Table A.I.—Continued**

| *Name* | *Symbol* | *Atomic Number* |
|---|---|---|
| Unnilpent-ium | Unp | 105 |
| Unnilquad-ium | Unq | 104 |
| Unnilsept-ium | Uns | 107 |
| Uran-ium | U | 92 |
| Vanad-ium | V | 23 |
| Xen/on- | Xe | 54 |
| Ytterb-ium | Yb | 70 |
| Yttr-ium | Y | 39 |
| Zinc- | Zn | 30 |
| Zircon-ium | Zr | 40 |

NOTE: The end of the longest stem for an element is indicated by a hyphen. A slash indicates the end of an alternate stem. If there is no hyphen within or at the end of the name for an element, that name is not used for the stem of that particular element.

**Proposed Names Not Accepted**
**(Accepted Name in Parentheses)**

Alabamine (Astatine)
Celtium (Hafnium)
Columbium (Niobium)
Florentium (Promethium)
Glucinium (Beryllium)
Illinium (Promethium)
Masurium (Technetium)
Virginium (Francium)

**Proposed Names Still the**
**Subject of Controversy**
**(Systematic Name in Parentheses)**

Hahnium (Unnilpentium)
Kurchatovium (Unnilquadium)
Nielsbohrium (Unnilpentium)
Rutherfordium (Unnilquadium)

**Table A.II. Numerical Terms Used as Multiplicative Prefixes**

| *Number* | *Type 1* | *Type 2* | *Type 3* | *Number (contd.)* | *Type 1 (contd.)* |
|---|---|---|---|---|---|
| 1 | mono[a] | | | 13 | trideca |
| 2 | di | bis | bi | | etc. |
| 3 | tri | tris | ter | 20 | icosa[b] |
| 4 | tetra | tetrakis | quater | 21 | henicosa[b] |
| 5 | penta | pentakis | quinque | 22 | docosa |
| 6 | hexa | hexakis | sexi | 23 | tricosa |
| 7 | hepta | etc. | septi | | etc. |
| 8 | octa | | octi | 30 | triaconta |
| 9 | nona | | novi | 31 | hentriaconta |
| 10 | deca | | deci | | etc. |
| 11 | undeca | | undeci | 40 | tetraconta |
| 12 | dodeca | | | | etc. |
| | | | | 100 | hecta |
| | | | | 101 | henhecta |
| | | | | | etc. |

NOTE: *See* refs. 32f and 78 for expanded listings. Types 1, 2, and 3 are defined in Section 2.2.3.

[a]The prefix mono is usually omitted except for emphasis or when confusion might arise from its absence.

[b]This spelling, as given in the 1970 IUPAC Inorganic Rules as an alternative to (hen)eicosa (*22w*), was also included in the 1979 edition of the IUPAC Organic Nomenclature Rules (*34r*). CAS, however, still retains the spelling (hen)eicosa.

**Table A.III. Descriptor Terms**

| *Term* | *Meaning* |
|---|---|
| *A* | counterclockwise sequence in chiral structure |
| *aci* | acid form |
| allo | closely related to |
| *anello* | *see* boron hydrides (Section 11.3.3) |
| anhydro | abstraction of water |
| *antiprismo* | eight atoms bound into a square antiprism |
| *arachno* | *see* boron hydrides (Section 11.3.3) |
| *asym (as)* | asymmetrical structure |
| *C* | clockwise sequence in chiral structure |
| *canasto* | *see* boron hydrides (Section 11.3.3) |
| *catena*/catena | chain structure (not part of parent name when italicized); linear polymer |
| *cis* | two groups occupying adjacent positions of a plane (on same side) or polyhedron |
| *closo* | *see* boron hydrides (Section 11.3.3) |
| *commo* | *see* boron hydrides (Section 11.3.3) |
| *cyclo*/cyclo | ring structure (not part of parent name when italicized) |
| *d* | rotates plane of polarized light to the right (dextro) |
| de | removal of |
| *dodecahedro* | eight atoms bound into a triangulated dodecahedron |
| *fac* | three groups occupying the corners of a triangular face of a polyhedron |
| *hexahedro* | five or eight atoms bound into a hexahedron |
| *hexaprismo* | twelve atoms bound into a hexagonal prism |
| *hypercloso* | *see* boron hydrides (Section 11.3.3) |
| *hypho* | *see* boron hydrides (Section 11.3.3) |
| hypo | central atom in lower oxidation state |
| *icosahedro* | twelve atoms bound into a triangulated icosahedron |
| *isoarachno* | *see* boron hydrides (Section 11.3.3) |
| *iso*/iso | isomer of indicated species (not part of parent name when italicized) |
| *isocloso* | *see* boron hydrides (Section 11.3.3) |
| *isonido* | *see* boron hydrides (Section 11.3.3) |
| *klado* | *see* boron hydrides (Section 11.3.3) |
| *l* | rotates plane of polarized light to the left (levo) |
| *mer* | three groups occupying three adjoining corners of a polyhedron, not all on the same face |
| meso/*meso* | intermediate hydrated form of oxo acid; also optically inactive because of internal compensation |
| meta | low hydrated form of oxo acid |
| *nido* | *see* boron hydrides (Section 11.3.3) |
| *octahedro* | six atoms bound into a triangulated octahedron |
| ortho | fully or highest stable hydrated form of oxo acid |
| para | higher hydrated form of oxo acid |
| *pentaprismo* | ten atoms bound into a pentagonal prism |
| per | central atom with highest (or a high) oxidation number; also oxygen with the oxidation number of –I |
| *pileo* | *see* boron hydrides (Section 11.3.3) |
| pyro | oxo acid formed from two molecules of the ortho oxo acid by the loss of one molecule of water |
| *quadro* | four atoms bound into a quadrangle |
| *R* | clockwise sequence in chiral structure |

Continued on next page.

**Table A.III.—Continued**

| *Term* | *Meaning* |
|---|---|
| *rac* | racemic (frequently designated *dl* instead of *rac*) |
| *S* | counterclockwise sequence in chiral structure |
| super | high proportion of |
| *sym* (s) | symmetrical structure |
| *tetrahedro* | four atoms bound into a tetrahedron |
| *trans* | two groups directly across a central atom or plane from each other |
| *triangulo* | three atoms bound into a triangle |
| *triprismo* | six atoms bound into a triangular prism |

NOTE: *See* ref. 32g for descriptor terms used in organic nomenclature.

**Table A.IV. Pearson Symbols for Bravais Lattices**

| *System*[a] | *Pearson Symbol* |
|---|---|
| Triclinic P | *aP* |
| Monoclinic P | *mP* |
| Monoclinic C | *mC* |
| Orthorhombic P | *oP* |
| Orthorhombic C | *oC* |
| Orthorhombic F | *oF* |
| Orthorhombic I | *oI* |
| Tetragonal P | *tP* |
| Tetragonal I | *tI* |
| Hexagonal P[b] | *hP* |
| Rhombohedral R | *hR* |
| Cubic P | *cP* |
| Cubic F | *cF* |
| Cubic I | *cI* |

NOTE: Adapted from Table I in ref. 47.

[a]P, C, F, I, and R stand for primitive, one-face-centered, all-face-centered, body-centered, and rhombohedral unit cells, respectively.

[b]Hexagonal P and trigonal P are identical.

**Table A.V. Trivial Names for Selected Parent Groups of Atoms Arranged by Symbol for Central Element**

| *Formula*[a] | *Name* | *Related Names*[b] |
|---|---|---|
| $AlO_2^-$ | Aluminate ion | |
| $AsH_pO_q^{n-}$ | Analogous to name for the phosphorus group with the same formula, except that $As(OH)_3$ may be named arsenious as well as arsenous acid | |
| $AtO_n^-$ | Analogous to name for the chlorine group with the same formula | |
| $BHO_2^-$ | Boronate ion | HB(O)(OH): boronic acid |
| $BH_2O^-$ | Borinate ion | $H_2B(OH)$: borinic acid |
| $BO_2^-$ | Metaborate ion | B(O)(OH): metaboric acid |
| $BO_3^{3-}$ | Borate ion | $B(OH)_3$: boric acid |
| $BeO_2^{2-}$ | Beryllate ion | |
| $BiH_pO_q^{n-}$ | Analogous to name for the phosphorus group with the same formula | |
| $BrO_n^-$ | Analogous to name for the chlorine group with the same formula | |
| CHO | | |
| $CHO^+$ | Formyl cation | |
| CHO | Formyl radical | Formyl (S) |
| $CHO_2^-$ | Formate ion | Formyloxy (S)<br>HC(O)(OH): formic acid |
| $CH_2N$ | | |
| $H_2C(=N)^+$ | Methyleneaminyl cation | |
| HC(=NH) | Formimidoyl radical | Formimidoyl (S)<br>Iminomethyl (S) |
| $H_2C(=N)^-$ | Methyleneaminyl anion<br>Methyleneamide ion | |
| | | $CH_2$=NH: methanimine<br>methylenimine<br>methyleneamine |
| $CH_2NO$ | | |
| $H_2NC(O)^+$ | Carbamoyl cation | |
| $H_2NC(O)$ | Carbamoyl radical | Carbamoyl (S)<br>Aminocarbonyl (S) |
| $HC(O)NH^-$ | Formylamide ion | Formylamino (S)<br>Formamido (S) |
| $CH_2NO_2$ | | |
| $H_2NC(O)_2^-$ | Carbamate ion | Carbamoyloxy (S)<br>$H_2NC(O)(OH)$: carbamic acid |
| $O_2NCH_2^-$ | Nitromethanide | Nitromethyl (S) |
| $CH_2N_2O$ | | |
| $H_2NC(O)N=$ | | Carbamoylimino (S)<br>(Aminocarbonyl)imino (S) |
| $(HN)C(O)(NH)^{2-}$ | Carbonyldiamide ion | Carbonyldiimino (S)<br>–NHC(O)NH–: ureylene (S)<br>$H_2NC(O)NH_2$: urea |
| $CH_3N_2O$ | | |
| $H_2NC(O)(NH)^-$ | Carbamoylamide ion | (Aminocarbonyl)amino (S)<br>$H_2NC(O)NH$–: ureido (S)<br>$H_2NC(O)NH_2$: urea |
| CN | | |
| $CN^-$ | Cyanide ion | Cyano (A and S)<br>HCN: hydrogen cyanide |
| $NC^-$ | Isocyanide ion | Isocyano (A and S)<br>HNC: hydrogen isocyanide |

**Table A.V.—Continued**

| Formula[a] | Name | Related Names[b] |
|---|---|---|
| CNO | | |
| $OCN^-$ | Cyanate ion | HOCN: cyanic acid |
| $NCO^-$ | Isocyanate ion | HNCO: isocyanic acid |
| $ONC^-$ | Fulminate ion | HONC: fulminic acid |
| $CN_2^{2-}$ | Methanetetrayldiamide ion<br>Methanediylidenediamide ion | Methanetetrayldinitrilo (S)<br>Methanediylidenedinitrilo (S)<br>$C(N)(NH_2)$: cyanamide |
| $CO_3^{2-}$ | Carbonate ion | Carbonylbis(oxy) (S)<br>$C(O)(OH)_2$: carbonic acid |
| ClO | | |
| $ClO^-$ | Hypochlorite ion | Chlorooxy (S)<br>Cl(OH): hypochlorous acid |
| OCl– | | Chlorosyl (S) |
| $ClO_2$ | | |
| $ClO_2^-$ | Chlorite ion | Chlorosyloxy (S)<br>Cl(O)(OH): chlorous acid |
| $O_2Cl$– | | Chloryl (S) |
| $ClO_3$ | | |
| $ClO_3^-$ | Chlorate ion | Chloryloxy (S)<br>$Cl(O)_2(OH)$: chloric acid |
| $O_3Cl$– | | Perchloryl (S) |
| $ClO_4^-$ | Perchlorate ion | Perchloryloxy (S)<br>$Cl(O)_3(OH)$: perchloric acid |
| $CrO_2^-$ | Chromite ion | Cr(O)(OH): chromous acid |
| $CrO_4^{2-}$ | Chromate ion | $Cr(O)_2(OH)_2$: chromic acid |
| FO | | |
| $FO^-$ | Hypofluorite ion | Fluorooxy (S)<br>F(OH): hypofluorous acid |
| OF– | | Fluorosyl (S) |
| $FeO_2^-$ | Ferrate ion | Fe(O)(OH): ferric hydroxide oxide |
| $FeO_2^{2-}$ | Ferrite ion | $Fe(OH)_2$: ferrous hydroxide |
| $GeO_3^{2-}$ | Germanate ion | $Ge(O)(OH)_2$: germanic acid |
| $IO_n^-$ | Analogous to name for the chlorine group | with the same formula |
| $IO_6^{5-}$ | Orthoperiodate ion | $I(O)(OH)_5$: orthoperiodic acid |
| $MnO_4^-$ | Permanganate ion | $Mn(O)_3(OH)$: permanganic acid |
| $MnO_4^{2-}$ | Manganate ion | $Mn(O)_2(OH)_2$: manganic acid |
| $MoO_4^{2-}$ | Molybdate ion | $Mo(O)_2(OH)_2$: molybdic acid |
| NO | | |
| ON– | | Nitroso (S) |
| $NO_2$ | | |
| $NO_2^-$ | Nitrite ion | Nitrosooxy (S)<br>N(O)(OH): nitrous acid |
| $O_2N$– | | Nitro (S) |
| $NO_2^{2-}$ | Nitroxylate(2–) ion | $N(OH)_2$: nitroxylic acid |
| $NO_3^-$ | Nitrate ion | Nitrooxy (S)<br>$N(O)_2(OH)$: Nitric acid |
| $NbO_3^-$ | Niobate ion | $Nb(O)_2(OH)$: niobic acid |
| $PHO_2$ | | |
| $P(H)(O)_2^{2-}$ | Phosphonite ion | $P(H)(OH)_2$: phosphonous acid |
| HOP(O)= | | Phosphinico (S) |
| $P(H)(O)_3^{2-}$ | Phosphonate ion | $P(H)(O)(OH)_2$: phosphonic acid<br>$-P(O)(OH)_2$: phosphono (S) |

Continued on next page.

**Table A.V.—Continued**

| *Formula*[a] | *Name* | *Related Names*[b] |
|---|---|---|
| $PH_2O$ | | |
| $P(H)_2(O)^-$<br>$H_2P(O)–$ | Phosphinite ion | $P(H)_2(OH)$: phosphinous acid<br>Phosphinyl (S)<br>$HP(O)=$: phosphinylidene (S)<br>$P(O)\equiv$: phosphinylidyne (S)<br>$OP–$: phosphoroso (S) |
| $P(H)_2(O)_2^-$ | Phosphinate ion | $P(H)_2(O)(OH)$: phosphinic acid<br>$O_2P–$: phospho (S) |
| $P(O)_3^{3-}$ | Phosphite ion | $P(OH)_3$: phosphorous acid |
| $P(O)_4^{3-}$ | Phosphate ion | $P(O)(OH)_3$: phosphoric acid |
| $PbO_3^{2-}$ | Plumbate ion | |
| $PoO_n^{2-}$ | Analogous to name for the sulfur group with the same formula | |
| $PtO_3^{2-}$ | Platinate ion | |
| $ReO_4^{n-}$ | Analogous to name for the manganese group with the same formula | |
| $RuO_4^-$ | Perruthenate ion | |
| $RuO_4^{2-}$ | Ruthenate ion | |
| SHO | | |
| $SHO^-$<br>$HOS–$ | Sulfenate ion | $HS(OH)$: sulfenic acid<br>Sulfeno (S) |
| $SHO_2$ | | |
| $SHO_2^-$<br>$HOS(O)–$ | Sulfinate ion | $HS(O)(OH)$: sulfinic acid<br>Sulfino (S) |
| $SHO_3$ | | |
| $SHO_3^-$<br>$HOS(O)_2–$ | Sulfonate ion | $HS(O)_2(OH)$: sulfonic acid<br>Sulfo (S) |
| SO | | |
| $SO^{2+}$ | Sulfinyl cation<br>Thionyl cation | |
| $–S(O)–$ | | Sulfinyl (S) |
| $SO_2$ | | |
| $SO_2^{2-}$<br>$–S(O)_2–$ | Sulfoxylate ion | $S(OH)_2$: sulfoxylic acid<br>Sulfonyl (S)<br>Sulfuryl (S) |
| $SO_3$ | | |
| $SO_3^{2-}$<br>$SO(O–)_2$<br>$–SO_3^-$ | Sulfite ion | $S(O)(OH)_2$: sulfurous acid<br>Sulfinylbis(oxy) (S)<br>Sulfonato (S) |
| $SO_4$ | | |
| $SO_4^{2-}$<br>$SO_2(O–)_2$ | Sulfate ion | $S(O)_2(OH)_2$: sulfuric acid<br>Sulfonylbis(oxy) (S) |
| $SbH_pO_q^{n-}$ | Analogous to name for the phosphorus group with the same formula | |
| $SeO_n^{2-}$ | Analogous to name for the sulfur group with the same formula | |
| $SiO_3^{2-}$ | Metasilicate ion | $Si(O)(OH)_2$: metasilicic acid |
| $SiO_4^{4-}$ | Silicate ion | $Si(OH)_4$: silicic acid |
| $SnO_2^{2-}$ | Stannite ion | $Sn(OH)_2$: stannous hydroxide |
| $SnO_3^{2-}$ | Stannate ion | $Sn(O)(OH)_2$: stannic acid |
| $TaO_3^-$ | Tantalate ion | $Ta(O)_2(OH)$: tantalic acid |
| $TcO_4^{n-}$ | Analogous to name for the manganese group with the same formula | |
| $TeO_n^{2-}$ | Analogous to name for the sulfur group with the same formula | |
| $TeO_6^{6-}$ | Orthotellurate ion | $Te(OH)_6$: orthotelluric acid |
| $TiO_3^{2-}$ | Titanate ion | $Ti(O)(OH)_2$: metatitanic acid |
| $TiO_4^{4-}$ | Orthotitanate ion | $Ti(OH)_4$: orthotitanic acid |

**Table A.V.—Continued**

| *Formula*[a] | *Name* | *Related Names*[b] |
|---|---|---|
| $UO_4^{2-}$ | Metauranate ion | $U(O)_2(OH)_2$: metauranic acid |
| $VO_3^-$ | Metavanadate ion | $V(O)_2(OH)$: metavanadic acid |
| $VO_4^{3-}$ | Orthovanadate ion | $V(O)(OH)_3$:orthovanadic acid |
| $WO_4^{2-}$ | Tungstate ion | $W(O)_2(OH)_2$: tungstic acid |
| $ZrO_p^{n-}$ | Analogous to name for the titanium group with the same formula | |

[a]The formulas for the ions are not intended to specify the location of the charge.

[b]The name of the species for additive nomenclature purposes is derived from the name for the species by the procedures described in Section 5.3.4 in most cases. Names marked (A) are to be used for additive nomenclature purposes, whereas names marked (S) are to be used for substitutive nomenclature purposes. More extensive compilations may be found in refs. 23 and 32h.

**Table A.VI. Structural Descriptors**

| Number of Atoms in CSU | Descriptor | Point Group | CEP Descriptor[a] |
|---|---|---|---|
| 3 | *triangulo* | $D_{3h}$ | |
| 4 | *quadro* | $D_{4h}$ | |
| 4 | *tetrahedro* | $T_d$ | [$T_d$-(13)-$\Delta^4$-*closo*] |
| 5 | *hexahedro* (triangulated) | $D_{3h}$ | $D_{3h}$-(131)-$\Delta^6$-*closo*] |
| 6 | *octahedro* | $O_h$ | [$O_h$-(141)-$\Delta^8$-*closo*] |
| 6 | *triprismo* | $D_{3h}$ | |
| 7 | | $D_{5h}$ | [$D_{5h}$-(151)-$\Delta^{10}$-*closo*] |
| 8 | *antiprismo* | $D_{4d}$ | |
| 8 | *hexahedro* (cube) | $O_h$ | |
| 8 | *dodecahedro* | $D_{2d}$ | [$D_{2d}$-(2222)-$\Delta^{12}$-*closo*] |
| 8 | | $D_{6h}$ | [$D_{6h}$-($1v^6 61 v^6$)-$\Delta^{12}$-*closo*] |
| 12 | *icosahedro* | $I_h$ | [$I_h$-(1551)-$\Delta^{20}$-*closo*] |

[a]The CEP descriptors (*see* Section 11.3.4) apply only to triangulated polyhedrons. Structural descriptors for other triangulated polyhedrons are given in Chart A.I.

**Table A.VII. Affixes (Prefixes or Infixes) of Functional Replacement Nomenclature**

| *Affix*[a] | *Replacement Operation* |
|---|---|
| Amido | –OH by $-NH_2$ |
| Azido[b] | –OH by $-N_3$ |
| Bromido (infix)<br>Bromo (prefix) | –OH by –Br |
| Chlorido (infix)<br>Chloro (prefix) | –OH by –Cl |
| Cyanatido (infix)<br>Cyanato (prefix) | –OH by –OCN |
| Cyanido (infix)<br>Cyano (prefix) | –OH by –CN |
| Dithioperoxo[b,c,d] | –O– by –SS– |
| Fluorido (infix)<br>Fluoro (prefix) | –OH by –F |
| Hydrazido | –OH by $-NHNH_2$ |
| Hydrazono[b] | =O by $=NNH_2$ |
| Imido | =O or –O– by =NH or –NH– |
| Iodido (infix)<br>Iodo (prefix) | –OH by –I |
| Isocyanatido (infix)<br>Isocyanato (prefix) | –OH by –NCO |
| Isothiocyanatido[d] (infix)<br>Isothiocyanato[d] (prefix) | –OH by –NCS |
| Nitrido | =O and –OH by ≡N |
| Peroxo | –O– by –OO– |
| Seleno | =O or –O– by =Se or –Se– |
| Telluro | =O or –O– by =Te or –Te– |
| Thio | =O or –O– by =S or –S– |
| Thiocyanatido[d] (infix)<br>Thiocyanato[d] (prefix) | –OH by –SCN |
| Thioperoxo[b,d,e] | –O– by –OS– or –SO– |

[a]When the affixes are used as infixes, the final "o" of the affix is generally elided when followed by "a", "i", or "o".

[b]This affix is not listed in Section D (tentative) of the 1979 IUPAC Organic Rules (ref. 34e), but it is used in CA index nomenclature (ref. 32i).

[c]Enclosing this affix in parentheses assists greatly in recognition of the structure.

[d]The selenium and tellurium analogs are named similarly by replacing "thio" with "seleno" or "telluro" as appropriate.

[e]This affix and the corresponding selenium and tellurium affixes do not describe the chalcogen atom sequence. For other mixed chalcogen affixes, such as selenothioperoxo, it is possible to imply the sequence by the order of the chalcogen prefixes, but this has yet to be codified in recommendations.

**Table A.VIII. Names for Mononuclear Hydrides with Standard Bonding Numbers Toward Hydrogen**

| *Formula* | *Name* | *Standard Bonding Number* | *Formula* | *Name* | *Standard Bonding Number* |
|---|---|---|---|---|---|
| $BH_3$ | borane | 3 | $SH_2$ | sulfane<br>hydrogen sulfide[a] | 2 |
| $CH_4$ | carbane<br>methane[a] | 4 | $SeH_2$ | selane[b]<br>hydrogen selenide[a] | 2 |
| $SiH_4$ | silane | 4 | $TeH_2$ | tellane[b]<br>hydrogen telluride[a] | 2 |
| $GeH_4$ | germane | 4 | | | |
| $SnH_4$ | stannane | 4 | $PoH_2$ | polane<br>hydrogen polonide[a] | 2 |
| $PbH_4$ | plumbane | 4 | FH | fluorane<br>hydrogen fluoride[a] | 1 |
| $NH_3$ | azane<br>ammonia[a] | 3 | ClH | chlorane<br>hydrogen chloride[a] | 1 |
| $PH_3$ | phosphane<br>phosphine[a] | 3 | BrH | bromane<br>hydrogen bromide[a] | 1 |
| $AsH_3$ | arsane<br>arsine[a] | 3 | IH | iodane<br>hydrogen iodide[a] | 1 |
| $SbH_3$ | stibane<br>stibine[a] | 3 | AtH | astatane<br>hydrogen astatide[a] | 1 |
| $BiH_3$ | bismuthane[b]<br>bismuthine[a] | 3 | | | |
| $OH_2$ | oxidane[c]<br>water[a] | 2 | | | |

[a]Common trivial or semisystematic name.

[b]Bismane, selenane, and tellurane cannot be used because they are Hantzsch–Widman names for saturated six-membered rings with one heteroatom.

[c]Name recently introduced into provisional IUPAC Nomenclature Rules by CNOC because oxane is the Hantzsch–Widman name for a saturated six-membered heteromonocycle with one oxygen ring atom.

**Table A.IX. Hydrogen-Containing Inorganic Compounds with Nonhydride Names**

| *Formula* | *Name*[a] |
|---|---|
| $ClNH_2$ | chloramine (IUPAC)<br>chloramide (CAS) |
| $Cl_2NH$ | dichloramine (IUPAC)<br>chlorimide (CAS) |
| $NH_2OH$ | hydroxylamine |
| $H_2S(NH)$ | sulfimide (IUPAC)[b]<br>sulfilimine (CAS) |
| $H_2S(O)(NH)$ | sulfoximide (IUPAC)<br>sulfoximine (CAS) |
| $S(O)(NH)$ | thionyl imide |
| $S(NH)_2$ | sulfur diimide |
| $S(O)_2(NH)$ | sulfonyl imide (IUPAC)<br>sulfuryl imide (IUPAC)<br>sulfimide (CAS)[b] |
| $S(NH)_3$ | sulfur triimide |
| $H_2NC(S)SC(S)NH_2$ | thiuram monosulfide |
| $H_2NC(S)SSC(S)NH_2$ | thiuram disulfide |
| $H_3P(O)$ | phosphine oxide[c] |
| $(H_2C)NH(O)$ | nitrone |
| $C(O)(NH_2)_2$ | urea[d] |
| $C(NH)(NH_2)_2$ | guanidine |
| $H_2NC(NH)NHNHC(NH)NH_2$ | biguanidine |
| $H_2N[C(NH)NH]_nC(NH)NH_2$ | |
| $n = 1$ | biguanide |
| $n = 2$ | triguanide |
| $n = 3$, etc. | tetra, etc., guanide |
| $C(NH)_2$ | carbodiimide |
| $H_2NN{=}CHN{=}NH$ | formazan |
| $H_2NC(O)NHNH_2$ | semicarbazide[d] |
| $HOC(NH)NHNH_2$ | isosemicarbazide[d] |
| $H_2NNHC(O)NHNH_2$ | carbonohydrazide[d] |
| $H_2NNHC(NNH_2)OH$ | isocarbonohydrazide[d] |
| $HN{=}NC(O)NHNH_2$ | carbazone[d] |
| $HN{=}NC(O)N{=}NH$ | carbodiazone[d] |
| $H_2N[C(O)NH]_nC(O)NH_2$ | |
| $n = 1$ | biuret[d] |
| $n = 2$ | triuret[d] |
| $n = 3$, etc. | tetra, etc., uret[d] |

[a]Alternative more systematic names are possible for all these compounds.

[b]It is necessary to be sure whether sulfimide is being used with the IUPAC meaning of $H_2S(NH)$ or the CAS meaning of $S(O)_2(NH)$.

[c]Similar names for $H_3P(X)$ where X = S, Se, or Te and for the arsenic, antimony, and bismuth analogs.

[d]The corresponding sulfur, selenium, and tellurium compounds are named by using the prefixes thio-, seleno-, and telluro-.

**Table A.X. Replacement ("a") Morphemes in Decreasing Order of Seniority**

| Element | Bonding Number[a] | Morpheme | Element | Bonding Number[a] | Morpheme |
|---|---|---|---|---|---|
| F | 1 | fluora | W | | tungsta |
| Cl | 1 | chlora | V | | vanada |
| Br | 1 | broma | Nb | | nioba |
| I | 1 | ioda | Ta | | tantala |
| At | 1 | astata | Ti | | titana |
| O | 2 | oxa | Zr | | zircona |
| S | 2 | thia | Hf | | hafna |
| Se | 2 | selena | Sc | | scanda |
| Te | 2 | tellura | Y | | yttra |
| Po | 2 | polona | La | | lanthana |
| N | 3 | aza | Ce | | cera |
| P | 3 | phospha[b] | Pr | | praseodyma |
| As | 3 | arsa[b] | Nd | | neodyma |
| Sb | 3 | stiba[b] | Pm | | prometha |
| Bi | 3 | bisma | Sm | | samara |
| C | 4 | carba | Eu | | europa |
| Si | 4 | sila | Gd | | gadolina |
| Ge | 4 | germa | Tb | | terba |
| Sn | 4 | stanna | Dy | | dysprosa |
| Pb | 4 | plumba | Ho | | holma |
| B | 3 | bora | Er | | erba |
| Al | 3[c] | alumina | Tm | | thula |
| Ga | 3[c] | galla | Yb | | ytterba |
| In | 3[c] | inda | Lu | | luteta |
| Tl | 3[c] | thalla | Ac | | actina |
| Zn | 2[c] | zinca | Th | | thora |
| Cd | 2[c] | cadma | Pa | | protactina |
| Hg | 2 | mercura | U | | urana |
| Cu | | cupra | Np | | neptuna |
| Ag | | argenta | Pu | | plutona |
| Au | | aura | Am | | america |
| Ni | | nickela | Cm | | cura |
| Pd | | pallada | Bk | | berkela |
| Pt | | platina | Cf | | californa |
| Co | | cobalta | Es | | einsteina |
| Rh | | rhoda | Fm | | ferma |
| Ir | | irida | Md | | mendeleva |
| Fe | | ferra | No | | nobela |
| Ru | | ruthena | Lr | | lawrenca |
| Os | | osma | Be | 2[c] | berylla |
| Mn | | mangana | Mg | 2[c] | magnesa |
| Tc | | techneta | Ca | | calca |
| Re | | rhena | Sr | | stronta |
| Cr | | chroma | Ba | | bara |
| Mo | | molybda | Ra | | rada |

[a]When there is a number in this column, it is the accepted "standard bonding number" for the element in question and is implied for use of the morpheme in replacement nomenclature or in the Hantzsch–Widman system. If there is no number or the bonding number is not standard, the lambda convention (*see* Section 9.2) may be used in substitutive nomenclature.

[b]The 1979 edition of the Hantzsch–Widman system conventions states that phospha, arsa, and stiba should be replaced by phosphor, arsen, and antimon, respectively, when immediately followed by -in or -ine. This provision was dropped from the revised system (*58*).

[c]Provisional.

**Table A.XI. Suffixes for the Hantzsch–Widman System**

| *Ring Size* | *Saturated*[a] | *Unsaturated*[b] | *Ring Size* | *Saturated*[a] | *Unsaturated*[b] |
|---|---|---|---|---|---|
| 3 | irane[c] | irene[d] | 7 | epane | epine |
| 4 | etane[c] | ete | 8 | ocane | ocine |
| 5 | olane[c] | ole | 9 | onane | onine |
| 6A[e] | ane | ine[f] | 10 | ecane | ecine |
| 6B[g] | inane | ine[h] | | | |
| 6C[i] | inane | inine | | | |

NOTE: Based on 1982 recommendations found in ref. 58.

[a]Used when there are no double bonds in the ring.

[b]Used when there are a maximum number of noncumulative double bonds in the ring (at least one) and the ring atoms have the bonding numbers shown in Table A.X or indicated by the lambda convention (*see* Section 9.2).

[c]The suffixes "iridine", "etidene", and "olidine" are preferred for rings containing nitrogen (*58*).

[d]The suffix "irine" may be used for rings containing nitrogen only.

[e]Used when the least preferred heteroatom is O, S, Se, Te, Bi, or Hg.

[f]Oxine should not be used as an alternate name for pyran because oxine is still used as a trivial name for 8-quinolinol.

[g]Used when the least preferred heteroatom is N, Si, Ge, Sn, or Pb.

[h]Azine should not be used as an alternate name for pyridine because azine is a class name for =N–N= compounds.

[i]Used when the least preferred heteroatom is B, F, Cl, Br, I, P, As, or Sb.

**Chart A.1.**

**A. Structural Descriptors and Numbering[a] for Some Closed Polyboron Polyhedrons, $[B_nH_n]^{2-}$**

**1. $n = 4$**

[$T_d$-(1$v^3$3$v^3$)-$\Delta^4$-*closo*]
[$T_d$-(13)-$\Delta^4$-*closo*]
*closo-*

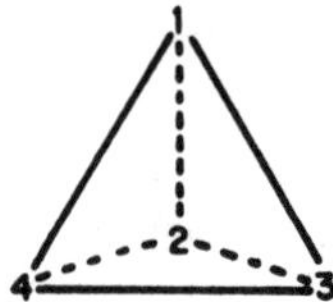


**2. $n = 5$**

[$D_{3h}$-(1$v^3$3$v^4$1$v^3$)-$\Delta^6$-*closo*]
[$D_{3h}$-(131)-$\Delta^6$-*closo*]
*closo-*

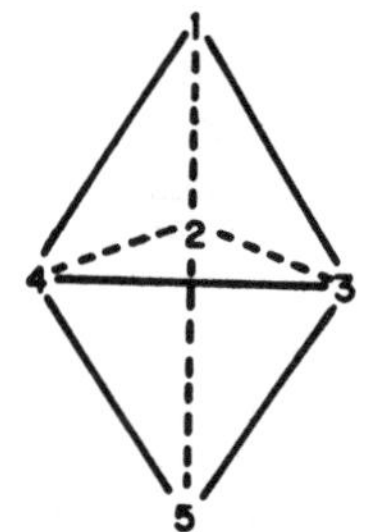


**3. $n = 6$**

[$O_h$-(1$v^4$4$v^4$1$v^4$)-$\Delta^8$-*closo*]
[$O_h$-(141)-$\Delta^8$-*closo*]
*closo-*

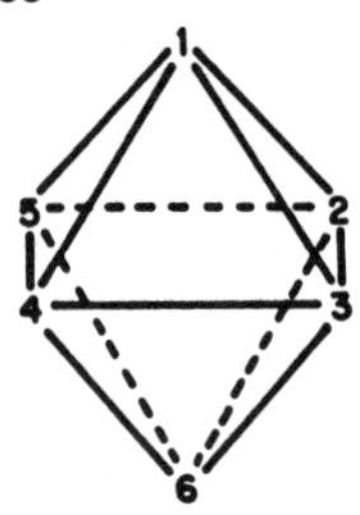


**4. $n = 7$**

[$D_{5h}$-(15$v^4$1)-$\Delta^{10}$-*closo*]
[$D_{5h}$-(151)-$\Delta^{10}$-*closo*]
*closo-*

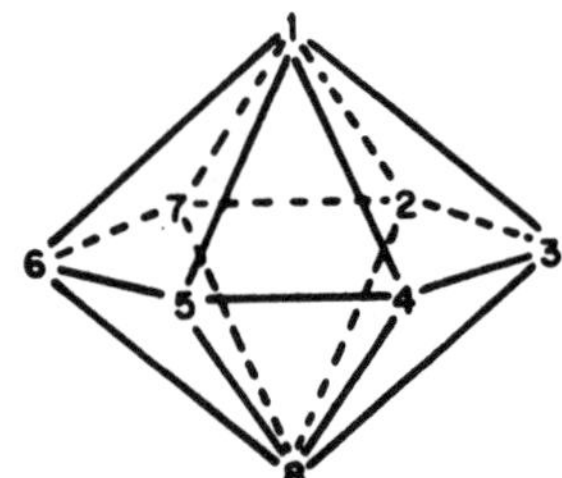


**5. $n = 8$**

[$D_{2d}$-(2$v^4$222$v^4$)-$\Delta^{12}$-*closo*]
[$D_{2d}$-(2222)-$\Delta^{12}$-*closo*]
*isocloso-*

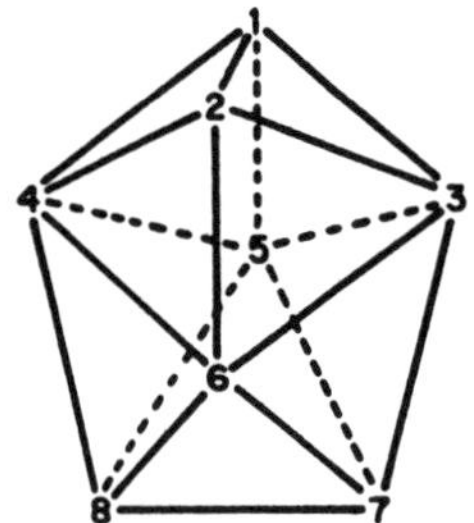


**6. $n = 8$**

[$D_{6h}$-(1$v^6$6$v^4$1$v^6$)-$\Delta^{12}$-*closo*]
[$D_{6h}$-(1$v^6$61$v^6$)-$\Delta^{12}$-*closo*]

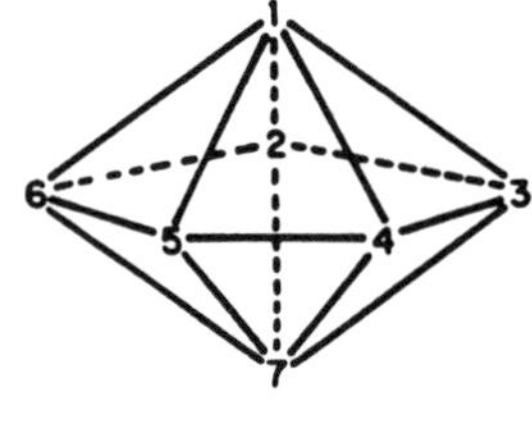


**7. $n = 9$**

[$D_{3h}$-(33$v^4$3)-$\Delta^{14}$-*closo*]
[$D_{3h}$-(333)-$\Delta^{14}$-*closo*]
*closo-*

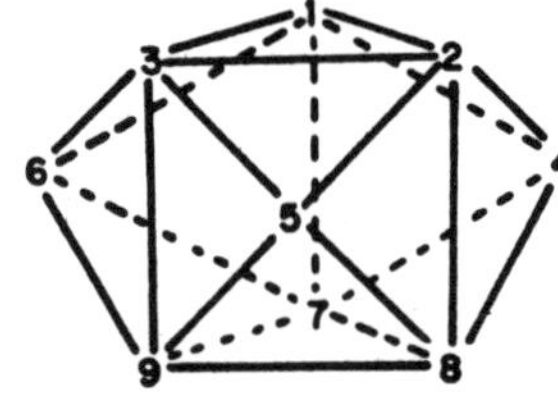


**8. $n = 9$**

[$C_{2v}$-(1$v^6$4$v^4$22)-$\Delta^{14}$-*closo*]

*isocloso-*

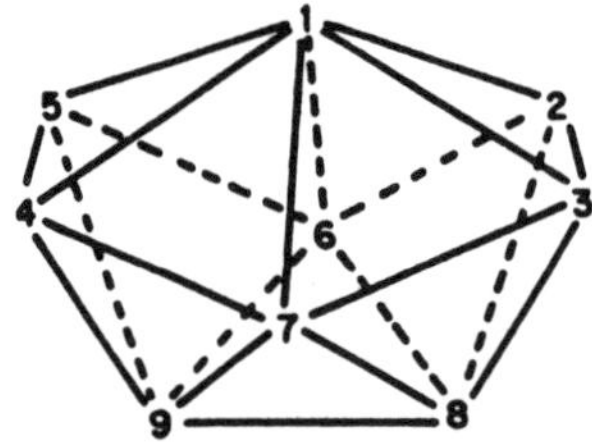


**9. $n = 9$**

[$C_{2v}$-(1$v^4$2$v^4$2$v^6$22$v^4$)-$\Delta^{14}$-*closo*]

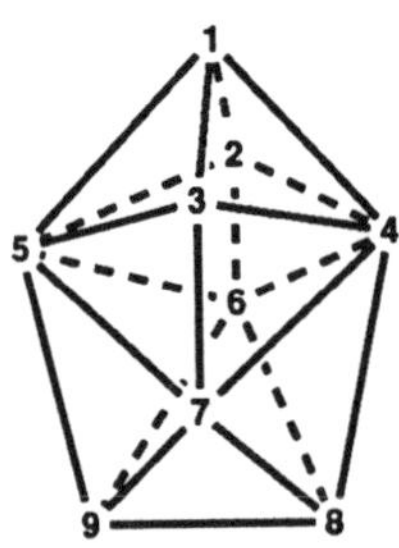

**Chart A.1—Continued**

**10. $n$ = 10**
[$D_{4d}$-(1$v^4$441$v^4$)-$\Delta^{16}$-*closo*]
[$D_{4d}$-(1441)-$\Delta^{16}$-*closo*]
*closo-*

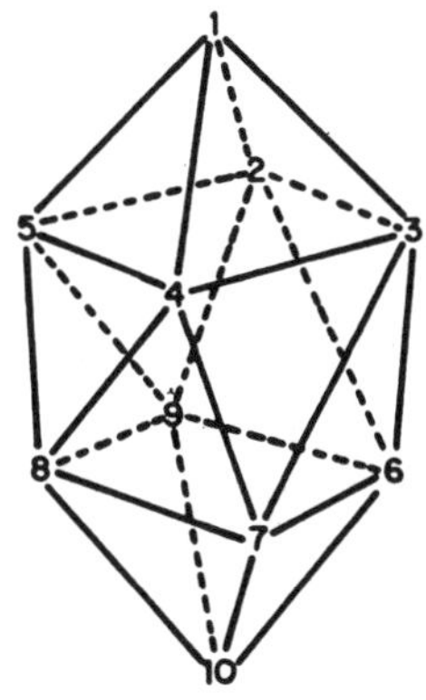

**11. $n$ = 10**
[$C_{3v}$-(1$v^6$3$v^4$33)-$\Delta^{16}$-*closo*]
*isocloso-*

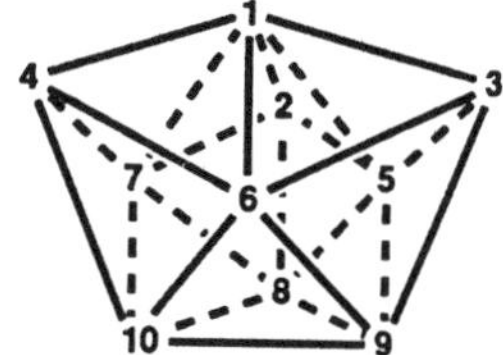

**12. $n$ = 11**
[$C_{2v}$-(1$v^6$2$v^4$422)-$\Delta^{18}$-*closo*]
*closo-* ("*isocloso-*")

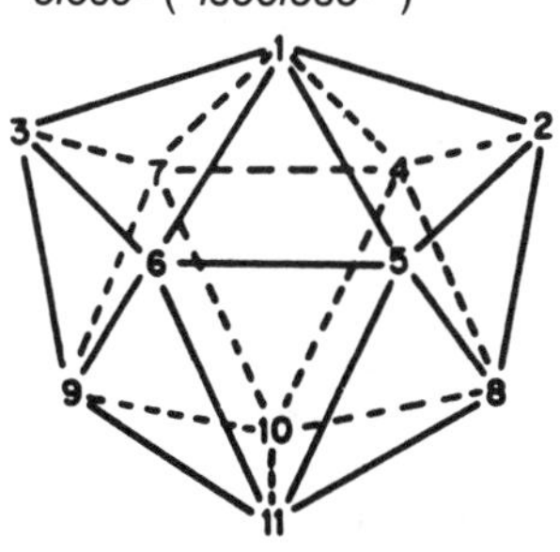

**13. $n$ = 11**
[$C_{2v}$-(1$v^4$422$v^6$2$v^4$)-$\Delta^{18}$-*closo*]

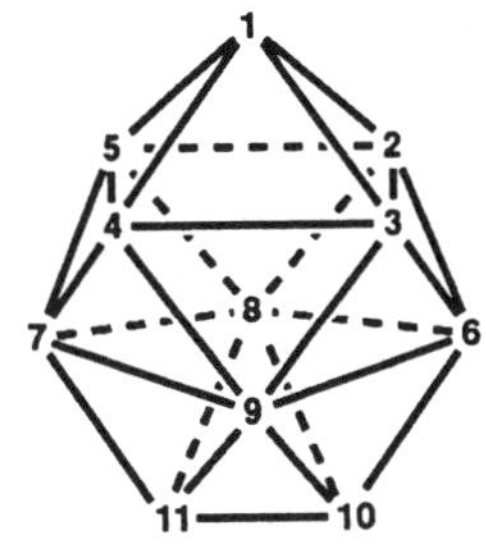

**14. $n$ = 12**
[$I_h$-(1551)-$\Delta^{20}$-*closo*]
*closo-*

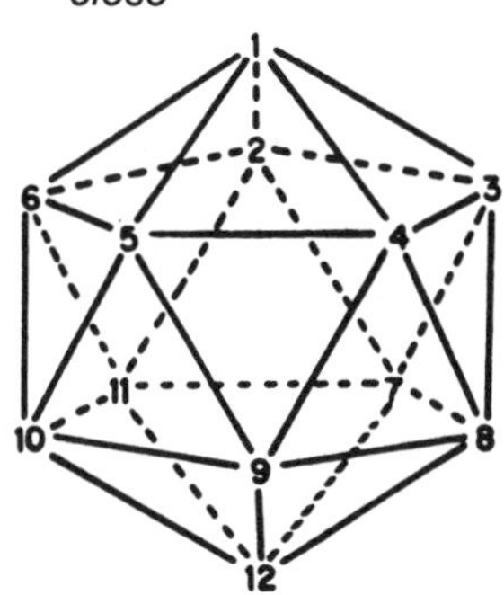

**15. $n$ = 13**
[$C_{2v}$-(1$v^4$22$v^6$422)-$\Delta^{22}$-*closo*]

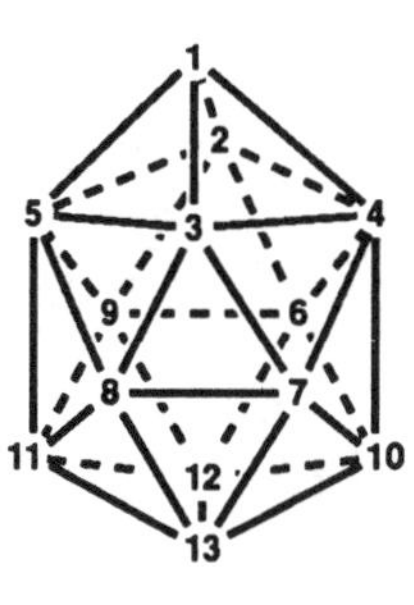

**16. $n$ = 14**
[$D_{6d}$-(1$v^6$661$v^6$)-$\Delta^{24}$-*closo*]

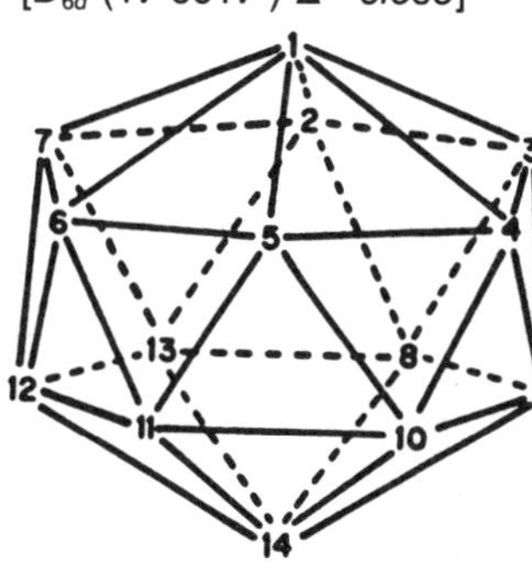

**17. $n$ = 14**
[$O_h$-(1$v^4$4$v^6$4$v^4$4$v^6$1$v^4$)-$\Delta^{24}$-*closo*]

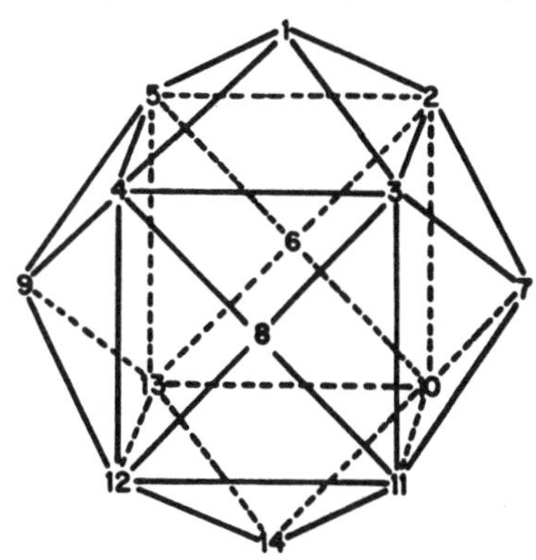

**18. $n$ = 15**
[$D_{3h}$-(333$v^6$33)-$\Delta^{26}$-*closo*]

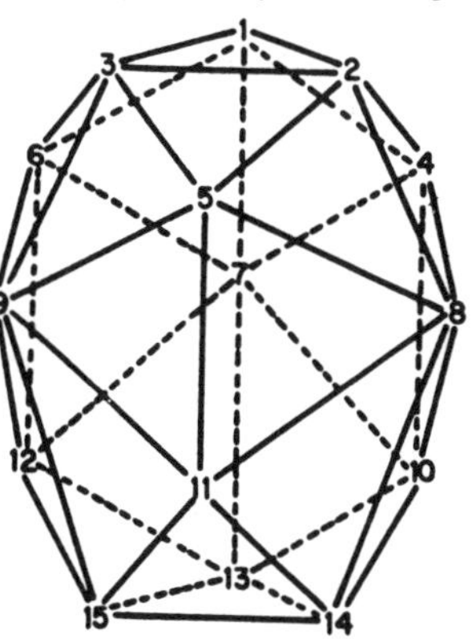

Continued on next page.

**Chart A.1—Continued**

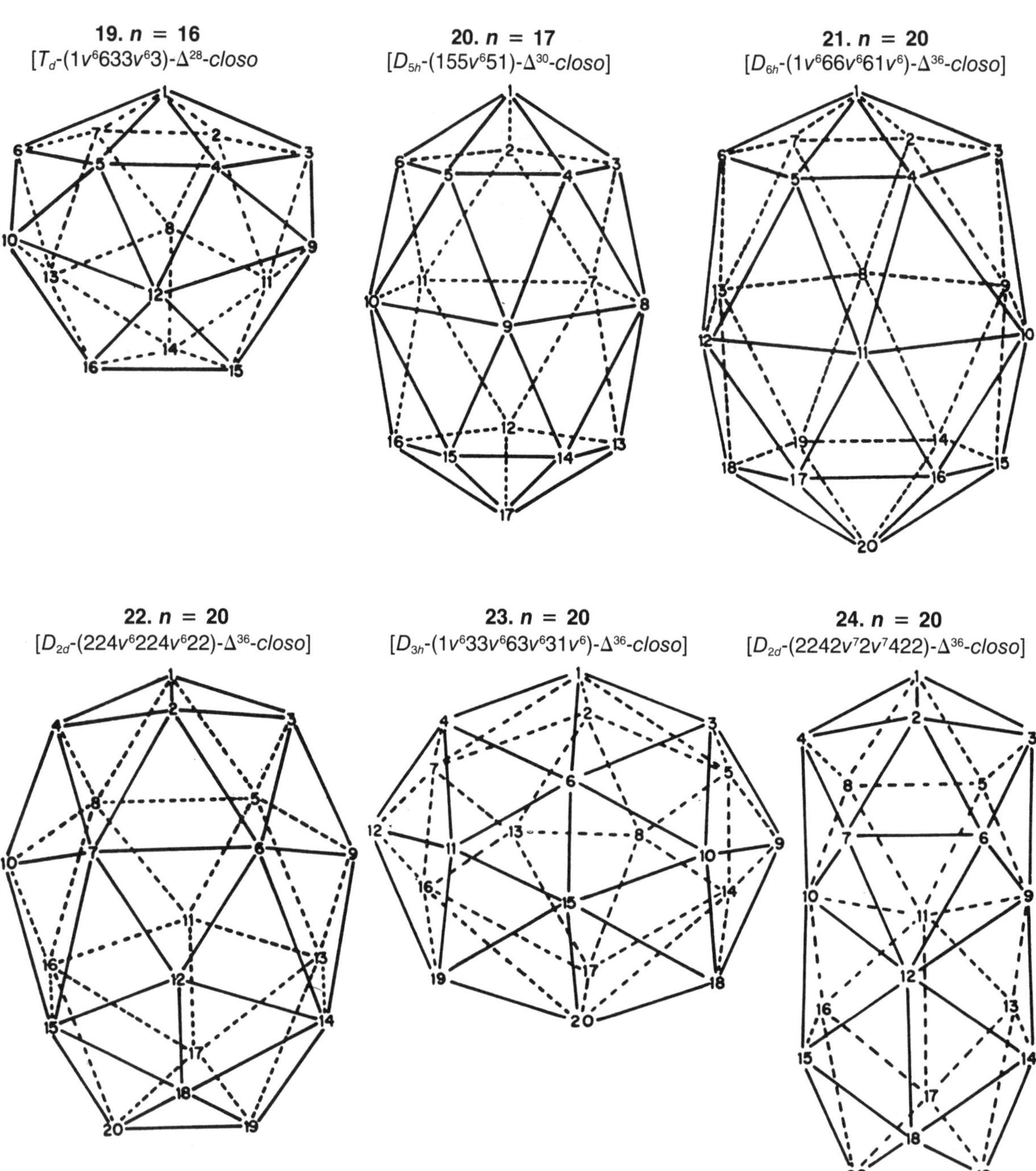


[a] The first descriptor listed for each structure is the full CEP descriptor (*see* Section 11.3.4), except that $v^n$ is omitted when $n = 5$; the second, if given, is the abbreviated CEP descriptor (*see* Section 11.3.4); and the third, if given, is the traditional descriptor (*see* Section 11.3.3). Only full descriptors have been given for $n > 12$. (Reproduced in part from ref. 53a. Copyright 1981, American Chemical Society.)

## Chart A.1—Continued

### B. Semisystematic Names and Numbering for Some Nonclosed Polyboron Structures

1. *nido*-pentaborane(9)

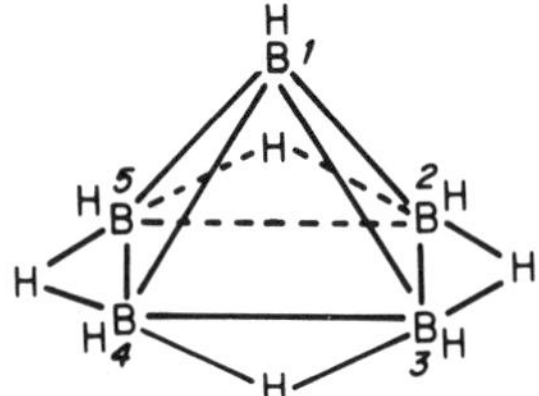

2. *nido*-hexaborane(10)

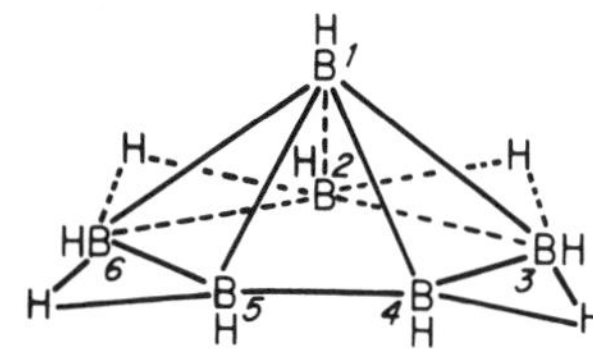

3. *nido*-decaborane(14)

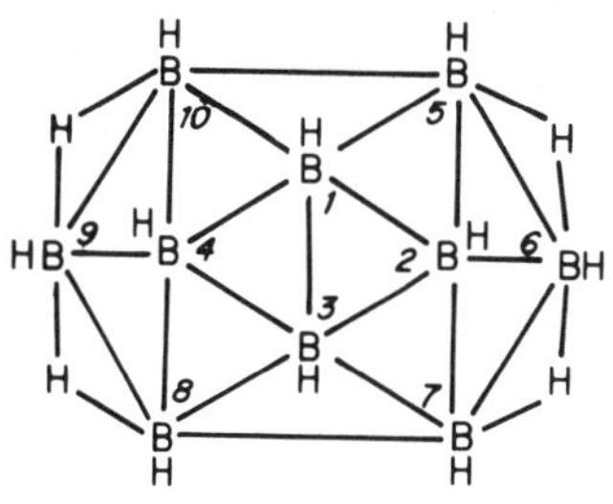

4. *nido*-undecaborane(15)

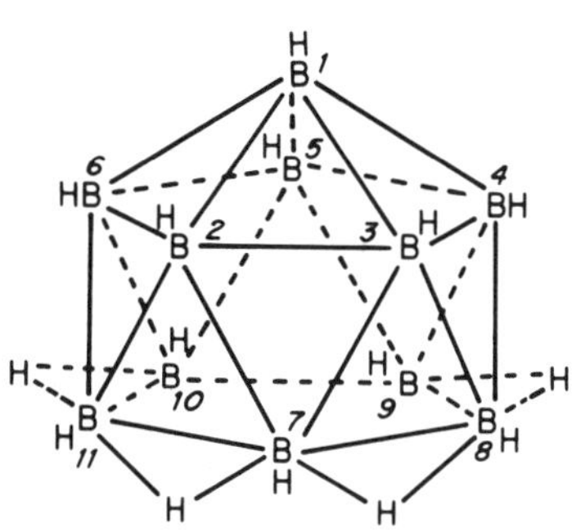

5. *isonido*-decabor polyhedron

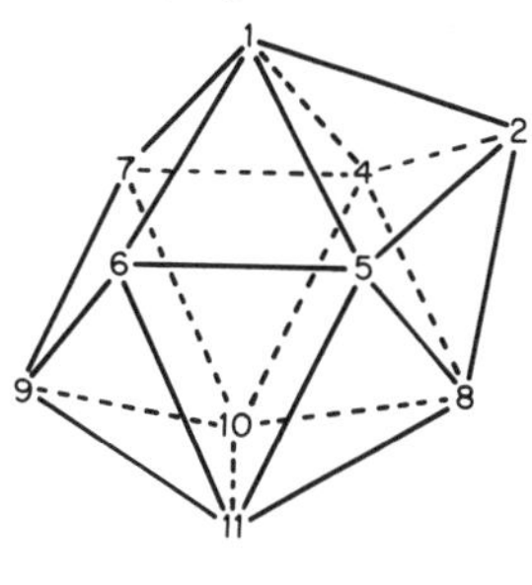

6. *nido*-nonabor polyhedron

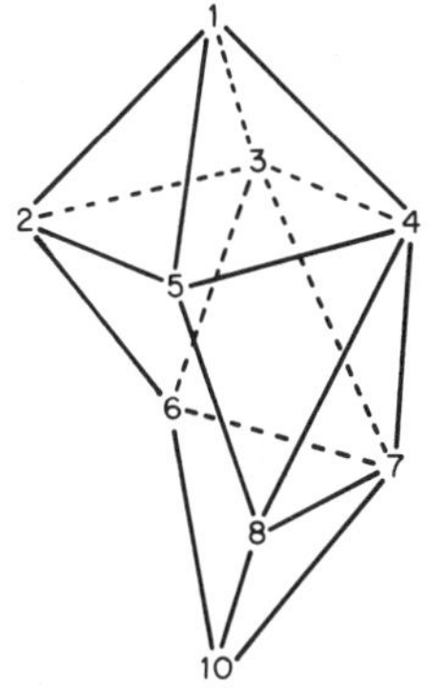

7. *arachno*-tetraborane(10)

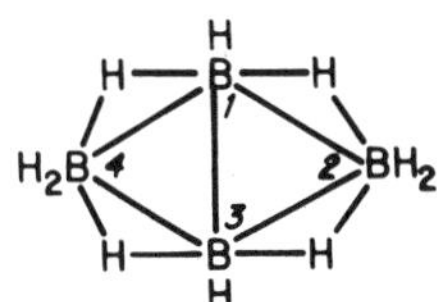

8. *arachno*-pentaborane(11)

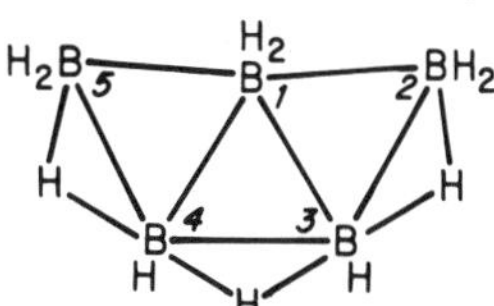

9. *arachno*-hexaborane(12)

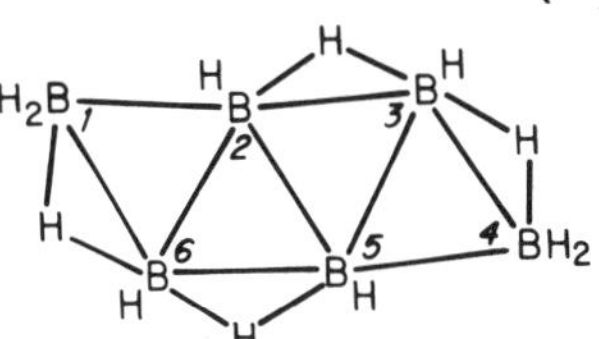

10. *arachno*-octaborane(14)

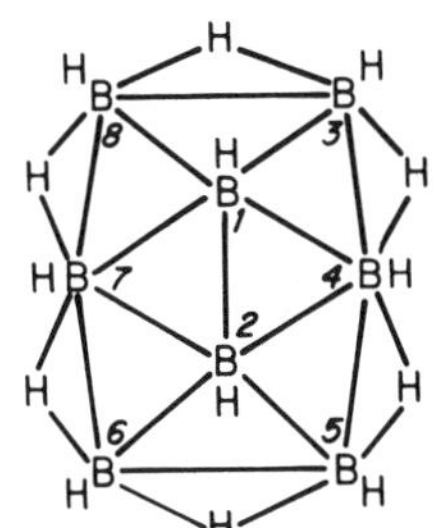

11. *arachno*-nonaborane(15)

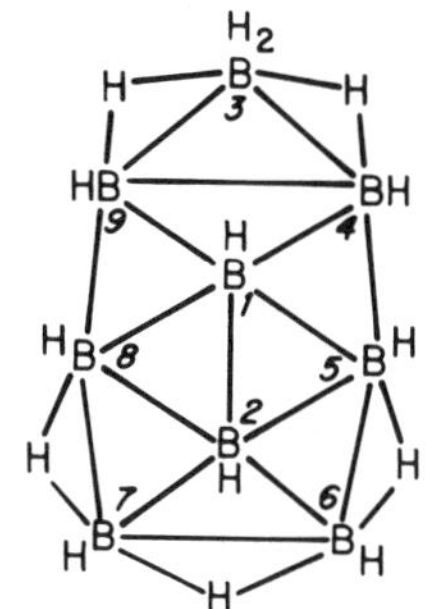

12. *isoarachno*-nonabor polyhedron

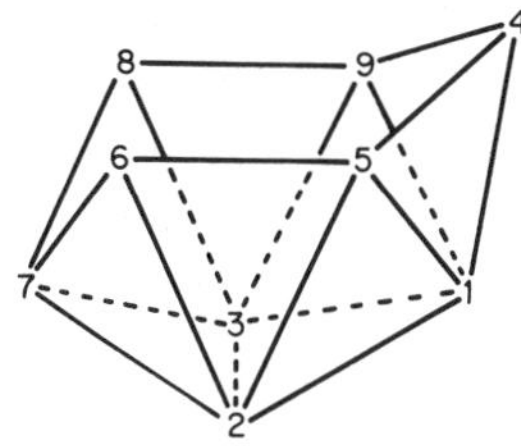

## Chart A.2

### Full Stereochemical Descriptors

The symmetry axes used in orienting the various structural figures are the symmetry axes for the underlying geometrical figures. In each case the order of seniority of the bonding atoms is alphabetical (i.e., a > b > c > d, etc.). Atoms lying above the reference plane are indicated by [a], etc., in the projection, those below by (a), etc. When there are more than two levels, multiple enclosing marks are used (e.g., ((a)) indicates an atom in the second plane below the reference plane). Eclipsed atoms are shown immediately following the eclipsing atom (e.g., [a]M(b) indicates that M is in the reference plane and that a is directly above it and b is directly below it).

### Bonding Number Two

▶ Linear (the bonding atoms and central atom form a straight line):

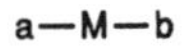

a—M—b

- Symmetry site term: *L*-2
- Orientation: View toward M along twofold axis
- Projection: Same as structural formula
- Configuration number: Not needed
- Chirality symbols: Only as needed for chiral side groups
- Stereochemical descriptor: *L*-2

▶ Angular (the bonding atoms and central atom do not form a straight line):

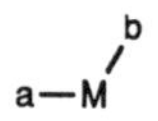

b
/
a—M

- Symmetry site term: *A*-2
- Orientation: View toward M on a line perpendicular to the plane formed by a, M, and b
- Projection: Same as structural formula
- Configuration number: Not needed
- Chirality symbols: Only as needed for chiral side groups
- Stereochemical descriptor: *A*-2

### Bonding Number Three

▶Trigonal Planar (the central atom lies in the plane formed by the bonding atoms):

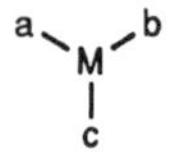

a ╲ M ╱ b
|
c

- Symmetry site term: *TP*-3
- Orientation: View toward M along threefold axis
- Projection: Same as structural formula
- Configuration number: Not needed
- Chirality symbols: Only as needed for chiral side groups
- Stereochemical descriptor: *TP*-3

▶ Trigonal Pyramidal (the central atom is not in the plane of the bonding atoms):

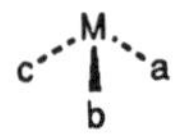

M
c ⋰ | ⋱ a
b

- Symmetry site term: *TPY*-3

**Chart A.2—Continued**

- Orientation: View plane of bonding atoms with M beneath the plane
- Projection:

- Configuration number: Not needed
- Chirality symbols: *R* or *S* when needed for the configuration about M; as needed for chiral side groups
- Stereochemical descriptor: *TPY*-3-*S* for projection shown

**Bonding Number Four**

▶ Tetrahedral:

- Symmetry site term: *T*-4
- Orientation: View from M along threefold axis toward least senior bonding atom
- Projection:

- Configuration number: Not needed
- Chirality symbols: *R* or *S* when needed for the configuration about M; as needed for chiral side groups
- Stereochemical descriptor: *T*-4-*R* for projection shown

▶ Square [Planar] (the bonding atoms form a square with the central atom in its plane):

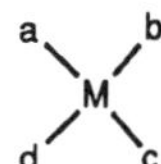


- Symmetry site term: *SP*-4
- Orientation: View toward M along fourfold axis
- Projection: Same as structural formula
- Configuration number: Seniority ranking number for the bonding atom trans to the most senior bonding atom
- Chirality symbols: Only as needed for chiral side groups
- Stereochemical descriptor: *SP*-4-3 for projection shown

**Bonding Number Five**

▶ Trigonal Bipyramidal:

- Symmetry site term: *TB*-5
- Orientation: View along threefold axis from more senior axial bonding atom toward M
- Projection:

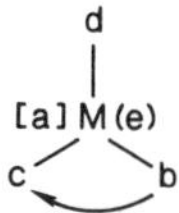


Continued on next page.

**Chart A.2—Continued**

- Configuration number: Seniority ranking numbers of the axial bonding atoms in numerical order
- Chirality symbols: *C* or *A* when needed for the configuration about M; as needed for chiral side groups
- Stereochemical descriptor: *TB*-5-15-*C* for projection shown

► Square Pyramidal:

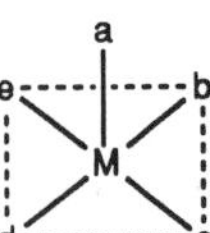


- Symmetry site term: *SPY*-5
- Orientation: View from the apex of the pyramid toward M
- Projection:

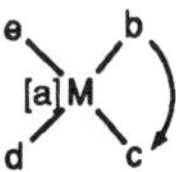


- Configuration number: Seniority ranking number of the apex bonding atom followed by the seniority ranking number of the bonding atom trans to the most senior bonding atom of the square in the projection
- Chirality symbols: *C* or *A* when needed for the configuration about M; as needed for chiral side groups
- Stereochemical descriptor: *SPY*-5-14-*C* for projection shown

**Bonding Number Six**

► Octahedral:

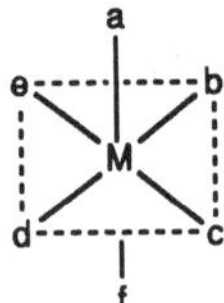


- Symmetry site term: *OC*-6
- Orientation: View from the most senior bonding atom toward M along the fourfold axis passing through the most senior bonding atom
- Projection:

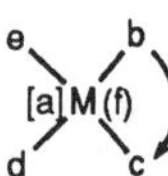


- Configuration number: Seniority ranking number of bonding atom trans to the most senior bonding atom followed by the seniority ranking number of the bonding atom trans to the most senior bonding atom of the square in the projection
- Chirality symbols: *C* or *A* when needed for the configuration about M except for the use of $\Delta$ or $\Lambda$ when at least two cis bidentate side groups are present (*see* Section 16.3.6); as needed for chiral side groups
- Stereochemical descriptor: *OC*-6-64-*C* for projection shown

► Trigonal Prismatic:

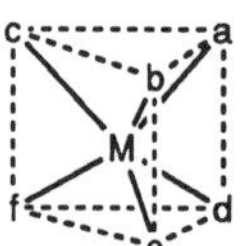


- Symmetry site term: *TP*-6
- Orientation: View along the threefold axis toward M from the triangular face containing the maximum number of bonding atoms of highest seniority (reference plane)

**Chart A.2—Continued**

- Projection:

- Configuration number: Seniority ranking numbers of the bonding atoms eclipsed by the bonding atoms in the plane of the projection (i.e., in the reference plane) cited in decreasing order of seniority of the eclipsing atoms
- Chirality symbols: *C* or *A* when needed for the configuration about M based on the direction of decreasing seniority of the eclipsed bonding atoms; as needed for chiral side groups
- Stereochemical descriptor: *TP*-6-456-*C* for projection shown

**Bonding Number Seven**

► Pentagonal Bipyramidal:

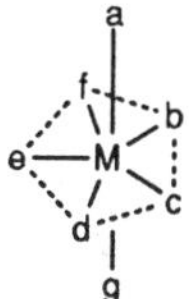


- Symmetry site term: *PB*-7
- Orientation: View toward M from the more senior bonding atom on the fivefold axis
- Projection:

- Configuration number: Seniority ranking numbers of the two axial bonding atoms in numerical order, hyphen, seniority ranking numbers of the remaining bonding atoms in sequence in the projection, starting with the most senior bonding atom and moving in the direction to give the lower configuration number
- Chirality symbols: *C* or *A* when needed for the configuration about M; as needed for chiral side groups
- Stereochemical descriptor: *PB*-7-17-23456-*C* for projection shown

► Mono(face-capped) Octahedral:

- Symmetry site term: *OCF*-7
- Orientation: View from bonding atom on threefold axis toward M
- Projection:

- Configuration number: Seniority ranking number of the axial bonding atom, hyphen, seniority ranking numbers of the remaining bonding atoms in sequence in the projection, starting with the most senior bonding atom in the plane adjacent to the axial bonding atom and moving in the direction to give the lower configuration number (moving in either direction leads to alternation of planes)
- Chirality symbols: *C* or *A* when needed for the configuration about M; as needed for chiral side groups
- Stereochemical descriptor: *OCF*-7-1-254736-*A* for projection shown

Continued on next page.

**Chart A.2—Continued**

► Mono(square-face-capped) Trigonal Prismatic:

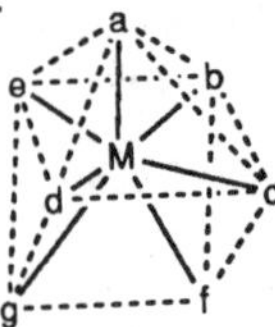


- Symmetry site term: *TPS*-7
- Orientation: View toward M from the axial bonding atom along the twofold axis
- Projection:

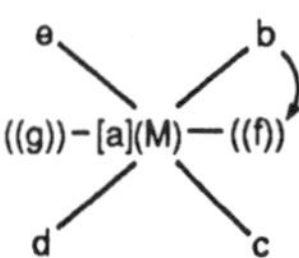


- Configuration number: Seniority ranking number of the axial bonding atom, hyphen, seniority ranking numbers of the remaining bonding atoms in sequence in the projection, starting with the most senior bonding atom in the plane adjacent to the axial bonding atom and moving in the direction so as to change planes
- Chirality symbols: *C* or *A* when needed for the configuration about M; as needed for chiral side groups
- Stereochemical descriptor: *TPS*-7-1-263475-*C* for projection shown

**Bonding Number Eight**

► Cubic:

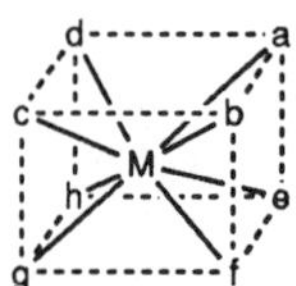


- Symmetry site term: *CU*-8
- Orientation: View toward M from a point on the fourfold axis above the face with the greatest number of most senior bonding atoms (the reference plane)
- Projection:

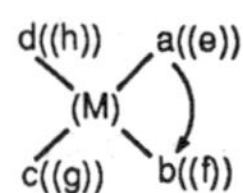


- Configuration number: Seniority ranking numbers of the bonding atoms are cited in sequence, starting with the most senior bonding atom in the reference plane and moving in the direction in the projection to give the lower configuration number. Seniority ranking numbers of eclipsed bonding atoms are cited immediately after those for the respective eclipsing atoms.
- Chirality symbols: *C* or *A* when needed for the configuration about M; as needed for chiral side groups
- Stereochemical descriptor: *CU*-8-15263748-*C* for projection shown

► Square Antiprismatic:

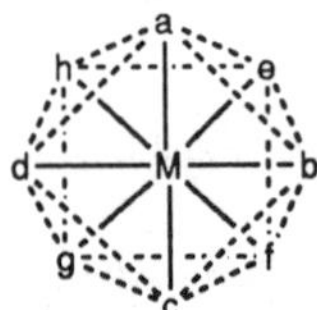


- Symmetry site term: *SA*-8

**Chart A.2—Continued**

- Orientation: View toward M from a point on the fourfold axis above the square face with the greater number of most senior bonding atoms (the reference plane)
- Projection:

```
        ((h))
   d      |     a
((g))——(M)——((e))
   c      |     b
        ((f))
```

- Configuration number: Seniority ranking numbers of the bonding atoms are cited in sequence, starting with the most senior atom in the reference plane and moving in the direction in the projection to give the lower configuration number (moving in either direction leads to alternation of planes)
- Chirality symbols: *C* or *A* when needed for the configuration about M; as needed for chiral side groups
- Stereochemical descriptor: *SA*-8-15263748-*C* for projection shown

► Dodecahedral:

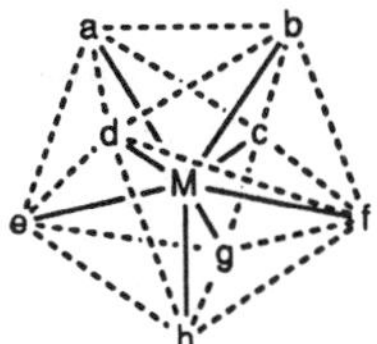


- Symmetry site term: *DD*-8
- Orientation: View toward M from a point on the twofold axis above the edge with the more senior bonding atoms
- Projection:

```
              (d)
               |
           ((((h))))
               |
(((f)))–b—((M))—a—(((e)))
               |
           ((((g))))
               |
              (c)
```

- Configuration number: Seniority ranking numbers of the bonding atoms are cited in sequence, starting with the more senior bonding atom in the top edge and moving in the direction in the projection to give the lower configuration number (moving in either direction leads to alternation of planes). Seniority numbers for eclipsed bonding atoms are cited immediately after their respective eclipsing bonding atoms.
- Chirality symbols: *C* or *A* when needed for the configuration about M; as needed for chiral side groups
- Stereochemical descriptor: *DD*-8-15372648-*C* for projection shown

► Hexagonal Bipyramidal:

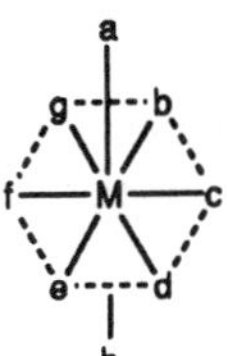


- Symmetry site term: *HB*-8
- Orientation: View toward M from more senior axial bonding atom
- Projection:

```
     g   b
f—[a]M(h)—c
     e   d
```

Continued on next page.

**Chart A.2—Continued**

- Configuration number: Seniority ranking numbers of the axial bonding atoms in numerical order, hyphen, seniority ranking numbers of the remaining bonding atoms in sequence in the projection, starting with the most senior bonding atom and moving in the direction in the projection to give the lower configuration number
- Chirality symbols: *C* or *A* when needed for the configuration about M; as needed for chiral side groups
- Stereochemical descriptor: *HB*-8-18-234567-*C* for projection shown

► *trans*-Bicapped Octahedral:

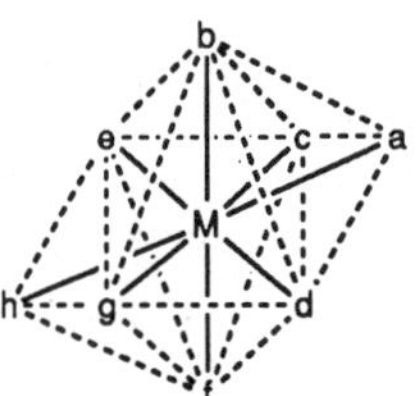

- Symmetry site term: *OCT*-8
- Orientation: View toward M from more senior axial bonding atom on threefold axis
- Projection:

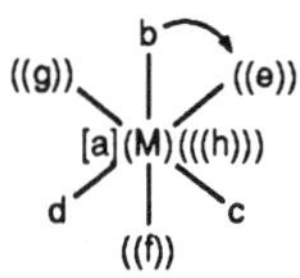

- Configuration number: Seniority ranking numbers of the axial bonding atoms in numerical order, hyphen, seniority ranking numbers of the remaining bonding atoms in sequence in the projection, starting with the most senior bonding atom in the plane adjacent to the more senior axial bonding atom and moving in the direction in the projection to give the lower configuration number (moving in either direction leads to alternation of planes)
- Chirality symbols: *C* or *A* when needed for the configuration about M; as needed for chiral side groups
- Stereochemical descriptor: *OCT*-8-18-253647-*C* for projection shown

► Bis(triangular-face-capped) Trigonal Prismatic:

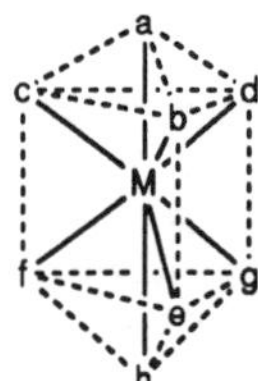

- Symmetry site term: *TPT*-8
- Orientation: View toward M from the more senior bonding atom on the threefold axis
- Projection:

- Configuration number: Seniority ranking numbers of the axial bonding atoms in numerical order, hyphen, seniority ranking numbers of the remaining bonding atoms in sequence in the projection, starting with the most senior bonding atom in the plane adjacent to the more senior axial bonding atom and moving in the direction in the projection to give the lower configuration number. Seniority ranking numbers for eclipsed bonding atoms are cited immediately after those for their respective eclipsing bonding atoms.
- Chirality symbols: *C* or *A* when needed for the configuration about M; as needed for chiral side groups
- Stereochemical descriptor: *TPT*-8-18-253647-*C* for projection shown

**Chart A.2—Continued**

► Bis(square-face-capped) Trigonal Prismatic:

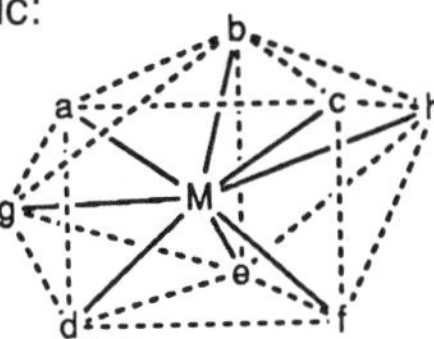


- Symmetry site term: *TPS*-8
- Orientation: View toward M along twofold axis from a point above the uncapped square face
- Projection:

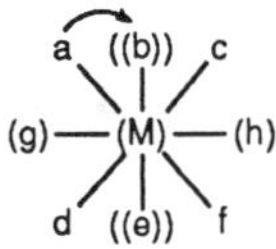


- Configuration number: Seniority ranking numbers of the bonding atoms in sequence starting with the most senior bonding atom in the uncapped square plane and moving in the direction in the projection to give the lower configuration number (moving in either direction leads to alternating planes)
- Chirality symbols: *C* or *A* when needed for the configuration about M; as needed for chiral side groups
- Stereochemical descriptor: *TPS*-8-12386547-*C* for projection shown

► Tris(square-face-capped) Trigonal Prismatic:

- Symmetry site term: *TPS*-9
- Orientation: View toward M from a point on the threefold axis above the triangular face containing the greater number of most senior bonding atoms (the reference plane)
- Projection:



- Configuration number: Seniority ranking numbers in sequence starting with the most senior bonding atom in the reference plane and moving in the direction in the projection to give the lower configuration number (moving in either direction leads to alternating planes). Seniority ranking numbers of eclipsed bonding atoms are cited immediately after those for their respective eclipsing bonding atoms
- Chirality symbols: *C* or *A* when needed for the configuration about M; as needed for chiral side groups
- Stereochemical descriptor: *TPS*-9-147258369-*C* for projection shown

► Heptagonal Bipyramid:

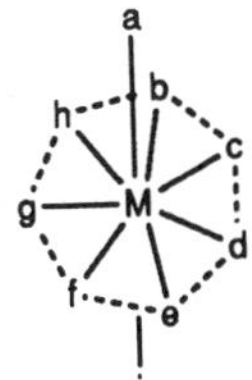


Continued on next page.

**Chart A.2—Continued**

- Symmetry site term: *HB*-9
- Orientation: View toward M along sevenfold axis from the more senior axial bonding atom
- Projection:

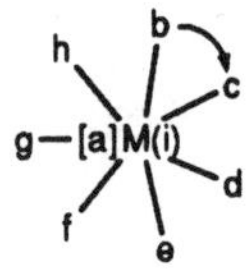

h, b, c, g—[a]M(i), d, f, e

- Configuration number: Seniority ranking numbers of the axial bonding atoms in numerical order, hyphen, seniority ranking numbers of the remaining bonding atoms in sequence in the projection, starting with the most senior bonding atom and moving in the direction to give the lower configuration number.
- Chirality symbols: *C* or *A* when needed for the configuration about M; as needed for chiral side groups
- Stereochemical descriptor: *HB*-9-19-2345678-*C* for projection shown

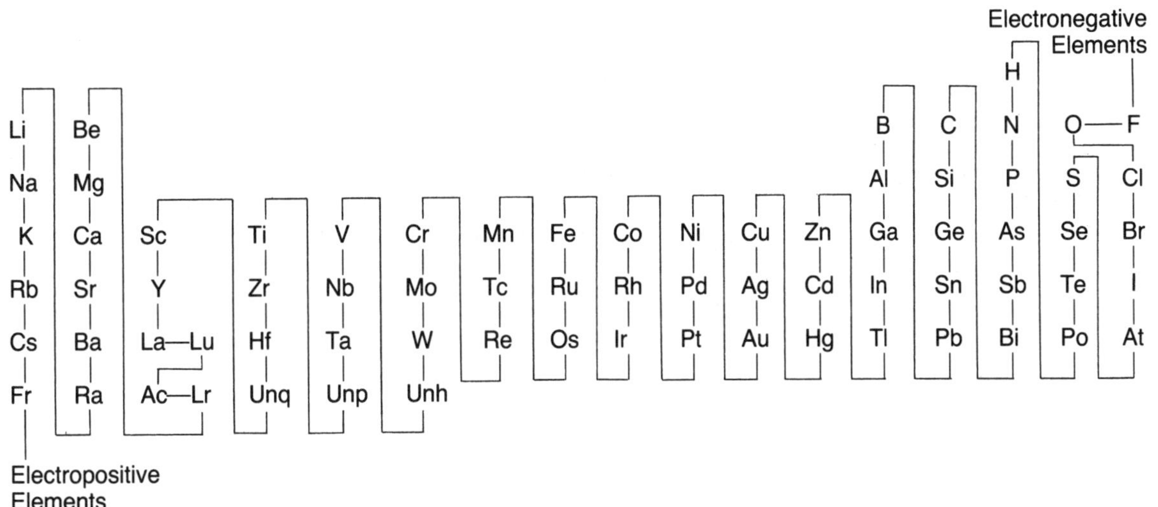


Figure A.1. Relative electronegativity order for binary nomenclature. This diagram presents a sequence of elements that is in accord with the ordering of the nonmetals in the 1970 IUPAC Inorganic Rules (*22a*) and also with the ordering of the metals and semimetals elsewhere in the same rules (*22c*). It is not identical with the latter ordering of the elements (*22c*) in the treatment of O, the inclusion of H, and the omission of the Group 18 elements.

| 1 | 2 | 3 | 4 | 5 | 6 | 7 | 8 | 9 | 10 | 11 | 12 | 13 | 14 | 15 | 16 | 17 | 18 |
|---|---|---|---|---|---|---|---|---|---|---|---|---|---|---|---|---|---|
| (IA | IIA | IIIB | IVB | VB | VIB | VIIB | |— | VIII | —| | IB | IIB | IIIA | IVA | VA | VIA | VIIA | VIIIA)[a] |
| (IA | IIA | IIIA | IVA | VA | VIA | VIIA | |— | VIIIA | —| | IB | IIB | IIIB | IVB | VB | VIB | VIIB | )[b] |
| H | | | | | | | | | | | | | | | | | He |
| Li | Be | | | | | | | | | | | B | C | N | O | F | Ne |
| Na | Mg | | | | | | | | | | | Al | Si | P | S | Cl | Ar |
| K | Ca | Sc | Ti | V | Cr | Mn | Fe | Co | Ni | Cu | Zn | Ga | Ge | As | Se | Br | Kr |
| Rb | Sr | Y | Zr | Nb | Mo | Tc | Ru | Rh | Pd | Ag | Cd | In | Sn | Sb | Te | I | Xe |
| Cs | Ba | La-[c] | Hf | Ta | W | Re | Os | Ir | Pt | Au | Hg | Tl | Pb | Bi | Po | At | Rn |
| Fr | Ra | Ac-[d] | Unq | Unp | Unh | | | | | | | | | | | | |

[a]Deming designation (*44*).
[b]Derived from IUPAC designation (*44*).
[c]Lanthanoids: La, Ce, Pr, Nd, Pm, Sm, Eu, Gd, Tb, Dy, Ho, Er, Tm, Yb, and Lu.
[d]Actinoids: Ac, Th, Pa, U, Np, Pu, Am, Cm, Bk, Cf, Es, Fm, Md, No, and Lr.

Figure A.2. Periodic table of the elements.

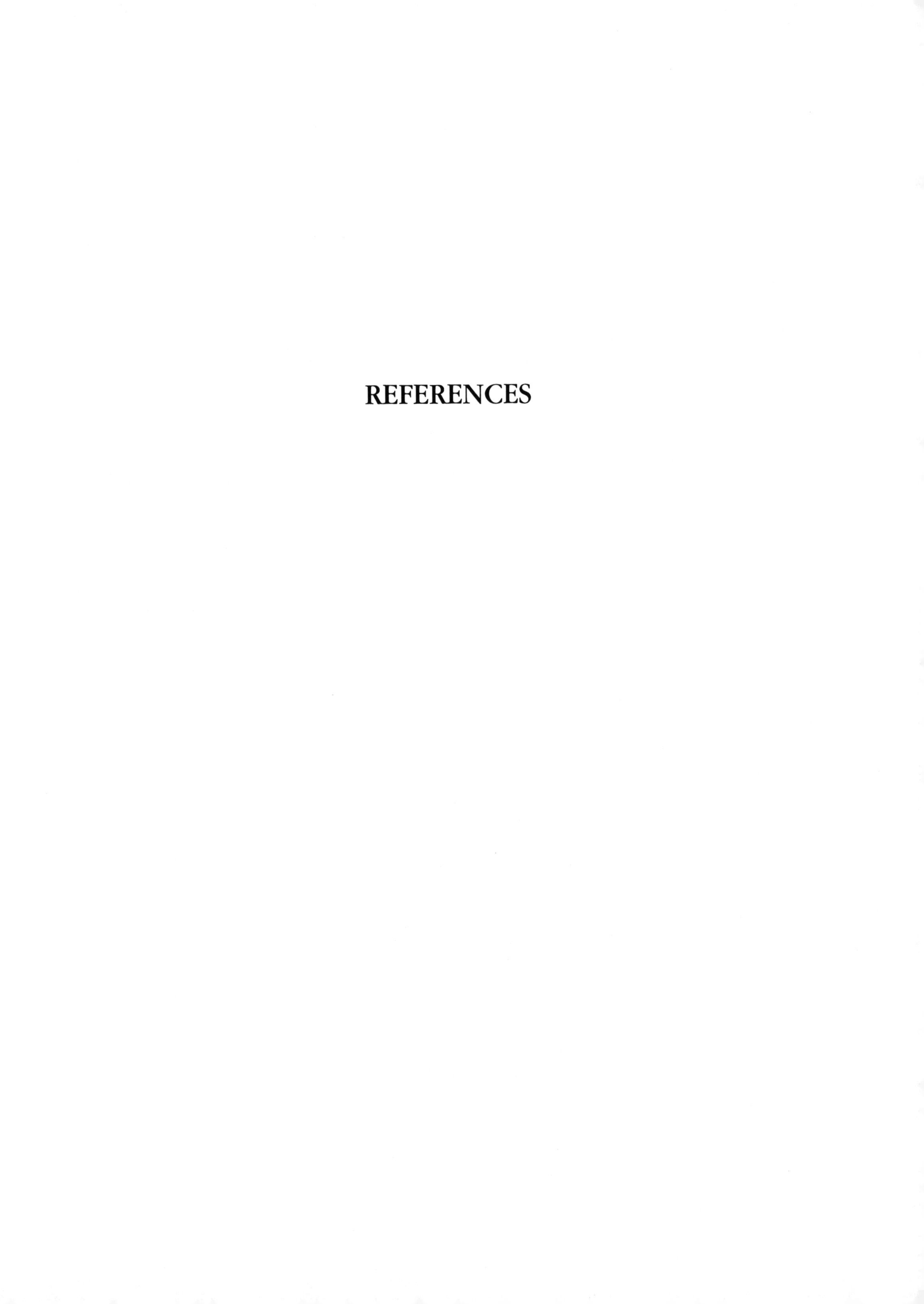

# REFERENCES

# References


1. *Nomenclature of Organic Compounds, Principles and Practice;* Fletcher, J. H., Dermer, O. C., and Fox, R. B., Eds.; Advances in Chemistry 126; American Chemical Society: Washington, DC, 1974; 337 pp.
2. *Webster's Third New International Dictionary of the English Language,* unabridged; Gove, P. B., Ed. in Chief; G. & C. Merriam: Springfield, MA, 1961.
3. Crossland, M. P. *Historical Studies in the Language of Chemistry,* 2nd ed.; Dover: New York, 1978; 406 pp.
4. Bergman, T. *Meditations de Systemate Fossilium Naturali;* J. Tofani: Florence, 1784; 125 pp.
5. Guyton de Morveau, L. B. "Sur les Denominations Chymiques, la necessite d'en perfectionner le systeme, & les regles pour y parvenir", *Obs. phys. hist. nat. arts* **1782,** *19,* 370–382; "De quelques critiques de la nomenclature des chimistes francais", *Ann. Chim. Paris* **1798,** *{1} 25,* 205–215.
6. Guyton de Morveau, L. B., Lavoisier, A. L., Bertholet, C. L., and de Fourcroy, A. F. *Methode de Nomenclature Chimique;* Cuchet: Paris, 1787; 314 pp; St. John, J. *Method of Chymical Nomenclature, English translation with adaptations to the English language;* G. Kearsley: London, 1788; 237 pp.
7. Loening, K. L. "Nomenclature", in *Kirk–Othmer Encyclopedia of Chemical Technology,* 3rd ed.; John Wiley and Sons: New York, 1981; Vol. 16, pp 28–46.
8. Lavoisier, A. L. *Traite Elementaire de Chimie,* 2 vols.; Cuchet: Paris, 1789; Kerr, R. *Elements of Chemistry,* English translation; William Creech: Edinburgh, 1790; 511 pp.
9. Berzelius, J. J. "Essai sur la Nomenclature Chimique", *J. Phys. Chim. Hist. Nat.* **1811,** *73,* 253–286.
10. Oersted, H. C. *Tentamen Nomenclaturae Chemicae Onmibus Linguis Scandinavico–Germanicis Communis;* Johannes: Copenhagen, 1814; 35 pp.
11. Werner, A. *Neuere Anschauungen auf dem Gebiete der anorganischen Chemie,* 3rd ed.; F. Vieweg und Sohn: Braunschweig, 1913; 419 pp.
12. Brauner, B. "Ueber die Stellung der Elemente der seltenen Erden im periodischen System von Mendelejeff", *Z. Anorg. Chem.* **1902,** *32,* 1–30 [p 10].
13. Chemical Society (London), Publication Committee. "Nomenclature and Notation", *J. Chem. Soc.* **1882,** *41,* 247–252.
14. American Chemical Society, Committee on Nomenclature and Notation. "Report", *J. Am. Chem. Soc.* **1886,** *8,* 116–125.
15. For the history of this association, *see* Ostwald, W. *Lebenslinien, Eine Selbstbiographie;* Klassing: Berlin, 1927; Vol. III, pp 262–286.
16. Proceedings of the Third Session of the Council of the International Association of Chemical Societies held at the Institut Solvay, Brussels, September 19–23, 1913 (from ref. 4 in *A History of the Nomenclature of Organic Chemistry,* Verkade, P. E.; English transl., Davies, S. G.; D. Reidel: Dordrecht, The Netherlands, 1985; Chapter IV, p 86).
17. International Union of Pure and Applied Chemistry. *Comptes Rendus de la Deuxieme Conference Internationale de la Chimie;* Brussels, June 27–30, 1921; pp 53–54.
18. Delepine, M. "Revision of Inorganic Chemical Nomenclature", *Chem. Weekblad* **1926,** *23,* 86–93; "Reforme de la Nomenclature de Chimie Minerale", *Bull. Soc. Chim. Fr.* **1928,** *43,* 289–300.
19. Jorissen, W. P., Bassett, H., Damiens, A., Fichter, F., and Remy, H. "Rules for Naming Inorganic Compounds, Report of the Committee of the International Union of Chemistry for the Reform of Inorganic Chemical Nomenclature, 1940", *J. Am. Chem. Soc.* **1941,** *63,* 889–897; cf. *Ber. Dtsch. Chem. Ges. A* **1940,** *73,* 53–70; *J. Chem. Soc.* **1940,** 1404–1415.

20. International Union of Pure and Applied Chemistry, Inorganic Chemistry Division, Commission on Nomenclature of Inorganic Chemistry. *Nomenclature of Inorganic Chemistry, Definitive Rules for Nomenclature of Inorganic Chemistry, 1957 Report;* Butterworths: London, 1959; 93 pp.
21. International Union of Pure and Applied Chemistry, Inorganic Chemistry Division, Commission on Nomenclature of Inorganic Chemistry. *Nomenclature of Inorganic Chemistry, Definitive Rules for Nomenclature of Inorganic Chemistry, 1957 Report;* American version with comments, *J. Am. Chem. Soc.* **1960,** *82,* 5523–5544.
22. International Union of Pure and Applied Chemistry, Inorganic Chemistry Division, Commission on Nomenclature of Inorganic Chemistry. *Nomenclature of Inorganic Chemistry,* 2nd ed., *Definitive Rules 1970;* Butterworths: London, 1971; 110 pp; (a) §2.161, p 14; (b) §2.22, p 15; (c) Table IV, p 104; (d) §1.12, p 10; (e) §1.15, p 10; (f) §1.21, pp 10–11 and Appendix, p 106; (g) §1.22, p 11; (h) §7.31, pp 41–44; (i) §7.25, p 41; (j) §7.513, pp 58–63; (k) §7.421, pp 49–50; (l) §4.14, p 27; (m) §5.24, p 34; (n) §5.22, p 33 and §5.23, p 34; (o) §5.34, p 35; (p) §4.2, pp 28–30; (q) §4.22, pp 29–30; (r) §6.2, p 36; (s) §8, pp 84–85; (t) §9.11, p 86; (u) §7.514, pp 63–64; (v) §7.81, p 77; (w) §0.31, p 6.
23. International Union of Pure and Applied Chemistry, Inorganic Chemistry Division, Commission on Nomenclature of Inorganic Chemistry. *How to Name an Inorganic Substance, A Guide to the Use of Nomenclature of Inorganic Chemistry: Definitive Rules 1970;* Pergamon: Oxford, 1977; 36 pp.
24. International Union of Pure and Applied Chemistry, Inorganic Chemistry Division, Commission on Nomenclature of Inorganic Chemistry. "Nomenclature of Inorganic Boron Compounds", *Pure Appl. Chem.* **1972,** *30,* 683–710.
25. International Union of Pure and Applied Chemistry, Inorganic Chemistry Division, Commission on Nomenclature of Inorganic Chemistry. "Recommendations for the Naming of Elements of Atomic Numbers Greater than 100", *Pure Appl. Chem.* **1979,** *51,* 381–384.
26. International Union of Pure and Applied Chemistry, Inorganic Chemistry Division, Commission on Nomenclature of Inorganic Chemistry. "Nomenclature of Inorganic Chemistry: II.1–Isotopically Modified Compounds (Recommendations 1981)", *Pure Appl. Chem.* **1981,** *53,* 1887–1900.
27. International Union of Pure and Applied Chemistry, Inorganic Chemistry Division, Commission on Nomenclature of Inorganic Chemistry. "Nomenclature of Inorganic Chemistry: II.2. The Nomenclature of Hydrides of Nitrogen and Derived Cations, Anions, and Ligands (Recommendations 1981)", *Pure Appl. Chem.* **1982,** *54,* 2545–2552.
28. International Union of Pure and Applied Chemistry, Macromolecular Division and Inorganic Chemistry Division, Commission on Macromolecular Nomenclature and Commission on Nomenclature of Inorganic Chemistry. "Nomenclature for Regular Single-Strand and Quasi Single-Strand Inorganic and Coordination Polymers (Recommendations 1984)", *Pure Appl. Chem.* **1985,** *57,* 149–168.
29. International Union of Pure and Applied Chemistry, Inorganic Chemistry Division, Commission on Nomenclature of Inorganic Chemistry. "Nomenclature of Polyanions (Recommendations 1987)", *Pure Appl. Chem.* **1987,** *59,* 1529–1548.
30. International Union of Pure and Applied Chemistry, Inorganic Chemistry Division, Commission on Nomenclature of Inorganic Chemistry. *Nomenclature of Inorganic Chemistry, Recommendations 1990,* Leigh, G. J., Ed.; Blackwell Scientific: Oxford, 1990; 289 pp.
31. International Union of Pure and Applied Chemistry. "Revised Procedure for Comment and Approval of IUPAC Recommendations on Nomenclature and Symbols" in "IUPAC Recommendations on Nomenclature and Symbols: Revision of a Procedure", *Chem. Int.* **1983,** *5,* 51–53.
32. American Chemical Society, Chemical Abstracts Service. "Chemical Substance Index Names", Appendix IV, *Chemical Abstracts Index Guide 1989;* American Chemical Society, Chemical Abstracts Service: Columbus, OH, 1989; pp 111 I–254 I; (a) §184, p 163 I and §187, p 166 I; (b) §183, pp 158 I–160 I and §197, pp 171 I–173 I; (c) §192, p 168 I; (d) §220, pp 210 I–213 I; (e) §203 III, pp 184 I–187 I; (f) §311, p 239 I; (g) §309, pp 238 I–239 I; (h) §197, pp 171 I and 173 I, §219, pp 208 I–210 I, and §294, pp 226 I–232 I; (i) §129, p 121 I and §294, p 226 I.
33. American Chemical Society, Chemical Abstracts Service. *Chemical Abstracts Index Guide 1989;* American Chemical Society, Chemical Abstracts Service: Columbus, OH, 1989; 2020 pp.
34. International Union of Pure and Applied Chemistry, Organic Chemistry Division, Commission on

Nomenclature of Organic Chemistry. *Nomenclature of Organic Chemistry, Sections A, B, C, D, E, F, and H,* 1979 ed.; Rigaudy, J., and Klesney, S. P., Eds.; Pergamon: Oxford, 1979; 559 pp; (a) §C-0.5, pp 118–123; (b) §C-83.3, p 139 and C-84.4, pp 140–141; (c) §C-0.15, pp 105–107; (d) §C-4.1, pp 189–192; (e) §D-5.0, pp 382–383; (f) §D-5.53, pp 397–398; (g) §C-661, pp 243–244; (h) §C-462.1, p 200; (i) §C-0.1, pp 85–112; (j) §D-4.4, pp 377–378; (k) §B-1, pp 53–55; (l) §B-1.2, p 54; (m) §C-0.6, pp 123–127 and §B-4, pp 68–70; (n) §D-7.63, pp 449–450; (o) §B-3, pp 64–68; (p) §H, pp 513–538; (q) §E-0, p 473; (r) §A-1, p 5.

35. Cahn, R. S., Ingold, C., and Prelog, V. "Specification of Molecular Chirality", *Angew. Chem. Int. Ed. Engl.* **1966,** *5,* 385–414 (errata, *ibid.*, 511); *Angew. Chem.* **1966,** *78,* 413–447.
36. Stock, A. "Einige Nomenklaturfragen der anorganischen Chemie", *Z. Angew. Chem.* **1919,** *32 I,* 373–374; "Bemerkung zum vorstehenden Aufsatz über Nomenklaturfragen der anorganischen Chemie" {Rosenheim, A. "Einige Nomenklaturfragen der anorganischen Chemie zu dem gleichnamigen Aufsatz von A. Stock", *Z. Angew. Chem.* **1920,** *33, Aufsatzteil, 78–79}, Z. Angew. Chem.* **1920,** *33, Aufsatzteil,* 79–80.
37. Ewens, R. V. G., and Bassett, H. "Inorganic Chemical Nomenclature—Suggestions Put Forward", *Chem. Ind.* **1949,** 131–139.
38. Hayek, E. "Eine neue Deutsche Fassung der Nomenklaturregeln fuer die anorganische Chemie", *Oesterr. Chem. Z.* **1975,** *76,* 2–3.
39. Klemm, W. "Some Considerations on the Nomenclature of Inorganic Chemistry", *Rev. Roum. Chim.* **1977,** *22,* 639–643.
40. Hyde, E. K., Hoffman, D. C., and Keller, O. L., Jr. "A History and Analysis of the Discovery of Elements 104 and 105", *Radiochim. Acta* **1987,** *42,* 57–102.
41. Flerov, G. N., and Zvara, I. "Chemical Elements of the Second Hundred"; Jairt Institute of Nuclear Research Dubna, USSR; Report JINR–D7–6013, 1971; 88 pp.
42. Harvey, B. G., Hermann, G., Hoff, R. W., Hoffman, D. C., Hyde, E. K., Katz, J. J., Keller, O. L., Jr., Lefort, M., and Seaborg, G. T. "Criteria for the Discovery of Chemical Elements", *Science* **1976,** *193,* 1271–1272.
43. "Transfermium Elements" (a report based on open notes of meetings of an ad hoc joint IUPAC/IUPAP working group), *Chem. Int.* **1989,** *11,* 180–182.
44. Fernelius, W. C., and Powell, W. H. "Confusion in the Periodic Table of the Elements", *J. Chem. Educ.* **1982,** *59,* 504–508.
45. American Chemical Society, Committee on Nomenclature. "Recommended Format for the Periodic Table of the Elements", *J. Chem. Educ.* **1984,** *61,* 136.
46. American Chemical Society, Chemical Abstracts Service. *Chemical Abstracts Eleventh Collective Index, Volumes 96–105, 1982–1986, General Subjects;* American Chemical Society, Chemical Abstracts Service, Columbus, OH, 1987; p 974 GS.
47. Pearson, W. B. *A Handbook of Lattice Spacings and Structures of Metals and Alloys;* Pergamon: Oxford, 1967; Vol. 2, p 2.
48. International Union of Pure and Applied Chemistry, Organic Chemistry Division, Commission on Physical Organic Chemistry. "Glossary of Terms Used in Physical Organic Chemistry (Recommendations 1982)", *Pure Appl. Chem.* **1983,** *55,* 1281–1371.
49. International Union of Pure and Applied Chemistry. *Compendium of Chemical Terminology, IUPAC Recommendations;* Gold, V., Loening, K. L., McNaught, A. D., and Sehmi, P., Compilers; Blackwell Scientific: Oxford, 1987; 456 pp.
50. International Union of Pure and Applied Chemistry, Physical Chemistry Division, Commission on Molecular Structure and Spectroscopy. "Recommendations for Symbolism and Nomenclature for Mass Spectroscopy", *Pure Appl. Chem.* **1978,** *50,* 65–73.
51. Sloan, T. E., and Busch, D. H. "Structural Coordination Compound Nomenclature. Designation of Coordination Sites for Ambidentate and Flexidentate Ligands", *Inorg. Chem.* **1978,** *17,* 2043–2047.
52. Cotton, F. A. "Proposed Nomenclature for Olefin–Metal and Other Organometallic Complexes", *J. Am. Chem. Soc.* **1968,** *90,* 6230–6232.
53. Casey, J. B., Evans, W. J., and Powell, W. H. (a) "A Descriptor System and Principles for Numbering Closed Boron Polyhedra with at Least One Rotational Symmetry Axis and One Symmetry Plane",

*Inorg. Chem.* **1981,** *20,* 1333–1341; (b) "A Descriptor System and Suggested Numbering Procedures for Closed Boron Polyhedra Belonging to $D_n$, $T$, and $C_s$ Symmetry Point Groups", *Inorg. Chem.* **1981,** *20,* 3556–3561.

54. International Union of Pure and Applied Chemistry, Macromolecular Division, Commission on Macromolecular Nomenclature. "Nomenclature of Regular Single-Strand Organic Polymers (Rules Approved 1975)", *Pure Appl. Chem.* **1976,** *48,* 373–385.
55. American Chemical Society. "Report of the ACS Nomenclature, Spelling, and Pronunciation Committee for the First Half of 1952, E. Organic Compounds Containing Phosphorus", *Chem. Eng. News* **1952,** *30,* 4515–4522.
56. International Union of Pure and Applied Chemistry, Inorganic Chemistry Division, Commission on Nomenclature of Inorganic Chemistry. "Nomenclature for Inorganic Chains and Rings" (in preparation) [*see* Powell, W. H., and Sloan, T. E., "Inorganic Ring Nomenclature: Past, Present, and Future", *Phosphorus, Sulfur Silicon Relat. Elem.* **1989,** *41,* 183–191].
57. International Union of Pure and Applied Chemistry, Organic Chemistry Division, Commission on Nomenclature of Organic Chemistry. "Treatment of Variable Valence in Organic Nomenclature (Lambda Convention) (Recommendations 1983)", *Pure Appl. Chem.* **1984,** *56,* 769–778.
58. International Union of Pure and Applied Chemistry, Organic Chemistry Division, Commission on Nomenclature of Organic Chemistry. "Revision of the Extended Hantzsch–Widman System of Nomenclature for Heteromonocycles (Recommendations 1982)", *Pure Appl. Chem.* **1983,** *55,* 409–416.
59. International Union of Pure and Applied Chemistry, Organic Chemistry Division, Commission on Nomenclature of Organic Chemistry. "Nomenclature for Cyclic Organic Compounds with Contiguous Formal Double Bonds (The δ-Convention) (Recommendations 1988)", *Pure Appl. Chem.* **1988,** *60,* 1395–1401.
60. Bould, J., Greenwood, N. N., Kennedy, J. D., and McDonald, W. S. "Quantitative *ortho*-Cycloboronation of *P*-Phenyl Groups in Metallaborane Chemistry and the Crystal and Molecular Structure of the Novel *iso-closo*-Ten-vertex Metallaborane [1,1,1-H($PPh_3$)($Ph_2P$-*ortho*-$C_6H_4$)-*iso-closo*-(1-$IrB_9H_8$-2-)]", *J. Chem. Soc., Chem. Commun.* **1982,** 465–467; Greenwood, N. N. "Metalloborane Cluster Compounds", *Pure Appl. Chem.* **1983,** *55,* 77–87.
61. Greenwood, N. N. "Some New Metallaboranes with Novel Cluster Features", *Nova Acta Leopold* **1985,** *59,* 291–304; Elrington, M., Greenwood, N. N., Kennedy, J. D., and Thornton-Pett, M. "Preparation and Nuclear Magnetic Resonance Properties of Eleven-vertex *closo*-Type Osmaundecaboranes and the X-Ray Crystal Structure of the *ortho*-Cycloboronated Compound [2,5-$(OEt)_2$-1-($PPh_3$)-1-(*o*-$Ph_2C_6H_4$)-*closo*-1-$OsB_{10}H_7$-3]", *J. Chem. Soc., Dalton Trans.* **1986,** 2277–2282.
62. Crook, J. E., Greenwood, N. N., Kennedy, J. D., and McDonald, W. S. "A Novel Oxidation Insertion of a Metal Centre into a Degraded *closo*-Borane Cluster; Crystal and Molecular Structure of the Ten-vertex *iso-nido* Cluster [{IrC(OH)$B_8H_6$(OMe)}($C_6H_4PPh_2$)($PPh_3$)]", *J.Chem. Soc., Chem. Commun.* **1981,** 933–934.
63. Rudolph, R. W., and Thompson, D. A. "Systematics in Boron Hydride Reactivities. Acceptable Valence Structures and Rearrangement in Unimolecular and Bimolecular Nucleophilic and Electrophilic Reactions", *Inorg. Chem.* **1974,** *13,* 2779–2782; Rudolph, R. W. "Boranes and Heteroboranes: A Paradigm for the Electron Requirements of Clusters?", *Acc. Chem. Res.* **1976,** *9,* 446–452; Williams, R. E. "Coordination Number Pattern Recognition Theory of Carborane Structures", *Adv. Inorg. Chem. Radiochem.* **1976,** *18,* 67–142 [p 81].
64. Remmel, R. J., Johnson, H. D., II, Jaworiwsky, I. S., and Shore, S. G. "Preparation and Nuclear Magnetic Resonance Studies of the Stereochemically Nonrigid Anions $B_4H_9^-$, $B_5H_{12}^-$, $B_6H_{11}^-$, and $B_7H_{12}^-$. Improved Syntheses of $B_5H_{11}$ and $B_6H_{12}$", *J. Am. Chem. Soc.* **1975,** *97,* 5395–5403.
65. Fratini, A. V., Sullivan, G. W., Denniston, M. L., Hertz, R. K., and Shore, S. G. "Static Structure of the Fluxional Molecule $B_5H_9[P(CH_3)_3]_2$, an Isoelectronic Analog of $B_5H_{11}^{2-}$", *J. Am. Chem. Soc.* **1974,** *96,* 3013–3015.
66. Mangion, M., Hertz, R. K., Denniston, M. L., Long, J. R., Clayton, W. R., and Shore, S. G. "Derivatives of the Hypho Class of Boron Hydrides. I. The Molecular Structure and Nuclear Magnetic Resonance Spectra of Stereochemically Nonrigid $B_6H_{10}[P(CH_3)_3]_2$", *J. Am. Chem. Soc.* **1976,** *98,* 449–453.

67. Parry, R. W., and Kodama, G. "Mechanisms and Processes for Conversion of Smaller Boranes to Larger Boranes or Borane Fragments", Final Report, 1980; 21 pp [from *Gov. Rep. Announc. Index* (U.S.A.) **1981,** *81,* 2184 (AD–A093 783); *Chem. Abstr.* **1981,** *95,* 107722a].
68. Greenwood, N. N. "Novel Cluster Interactions in Metallaboranes", in *Rings, Clusters, and Polymers of the Main Group Elements;* Cowley, A. H., Ed.; ACS Symposium Series 232; American Chemical Society: Washington, DC, 1983; pp 125–138.
69. Jung, C. W., Baker, R. T., Knobler, C. B., and Hawthorne, M. F. "Closo and Hyper-Closo Ten-vertex Ruthenacarboranes Containing Chelating Alkenylphosphine Ligands", *J. Am. Chem. Soc.* **1980,** *102,* 5782–5790.
70. Elrington, M., Greenwood, N. N., Kennedy, J. D., and Thornton-Pett, M. "Ruthenium(II) Complexes of *closo*-Dodecaboranyl Anions and the Molecular Structure of the *pileo* Thirteen-vertex Ruthenaborane [$(PPh_3)_2ClRuB_{12}H_{11}(NEt_3)$]", *J. Chem. Soc., Dalton Trans.* **1987,** 451–456.
71. Stumpf, W. "Nomenklatur, Verbindungstypen und Chemie der Carborane", *Chem. Z.* **1975,** *99,* 1–12; Stumpf, W. in *Themen zur Chemie des Bor;* Mollinger, H., Ed.; Dr. Alfred Huthig Verlag GmbH: Heidelberg, 1976; pp 85–119; *Gmelin Handbuch der anorganischen Chemie*, 8 Aufl.; Springer: Berlin, 1974; Erganzungswerk, Band 15, Borverbindungen 2, Kapitel 1, s. 1–138 [p 25].
72. Casey, J. B., Evans, W. J., and Powell, W. H. (a) "Structural Nomenclature for Polyboron Hydrides and Related Compounds. 1. Closed and Capped Polyhedral Structures", *Inorg. Chem.* **1983,** *22,* 2228–2235; (b) "2. Nonclosed Structures", *Inorg. Chem.* **1983,** *22,* 2236–2245; (c) "3. Linear *conjuncto-* Structures", *Inorg. Chem.* **1984,** *23,* 4132–4143.
73. Boocock, S. K., Greenwood, N. N., Kennedy, J. D., McDonald, W. S., and Staves, J. "The Chemistry of Isomeric Icosaboranes, $B_{20}H_{26}$. Molecular Structures and Physical Characterization of 2,2′-Bi(*nido*-decaboranyl) and 2,6′-Bi(*nido*-decaboranyl)", *J. Chem. Soc., Dalton Trans.* **1980,** 790–796.
74. International Union of Pure and Applied Chemistry, Physical Chemistry Division, Commission on Colloid and Surface Chemistry. "Chemical Nomenclature and Formulation of Compositions of Synthetic and Natural Zeolites", *Pure Appl. Chem.* **1979,** *51,* 1091–1100.
75. International Union of Pure and Applied Chemistry, Inorganic Chemistry Division, Commission on Atomic Weights and Isotopic Abundances. "Atomic Weights of the Elements 1987", *Pure Appl. Chem.* **1988,** *60,* 841–854.
76. Boughton, W. A. "Naming Hydrogen Isotopes", *Science* **1934,** *79,* 159–160; Crane, E. J. "Nomenclature of the Hydrogen Isotopes and Their Compounds", *Science* **1934,** *80,* 86–89; American Chemical Society, "Report of Committee on Nomenclature, Spelling, and Pronunciation, Nomenclature of Hydrogen Isotopes and their Compounds", *Ind. Eng. Chem. (News Ed.)* **1935,** *13,* 200–201.
77. Brown, M. F., Cook, B. R., and Sloan, T. E. "Stereochemical Notation in Coordination Chemistry"; (a) "Mononuclear Complexes", *Inorg. Chem.* **1975,** *14,* 1273–1278; (b) "Mononuclear Complexes of Coordination Numbers Seven, Eight, and Nine", *Inorg. Chem.* **1978,** *17,* 1563–1568.
78. International Union of Pure and Applied Chemistry, Organic Chemistry Division, Commission on Nomenclature of Organic Chemistry. "Extension of Rules A–1.1 and A–2.5 Concerning Numerical Terms Used in Organic Chemical Nomenclature (Recommendations 1986)", *Pure Appl. Chem.* **1986,** *58,* 1693–1696.

# Index

# Index

## C

## O



## P

## R

## T



## U



## V

W



Z



*Copy Editing, Production, Indexing: Colleen P. Stamm*
*Acquisition: Robin Giroux*
*Cover Design: Lori Seskin-Newman*

*Typeset by Techna Type, Inc., York, PA*
*Printed and bound by Maple Press, York, PA*